Building Materials Technology

Other McGraw-Hill Books of Interest

BEALL • *Masonry Design and Detailing*
BREYER • *Design of Wood Structures*
BROCKENBROUGH AND MERRITT • *Structural Steel Designer's Handbook*
FAHERTY • *Wood Engineering and Construction Handbook*
GAYLORD • *Structural Engineering Handbook*
MERRITT • *Standard Handbook for Civil Engineers*
MERRITT AND RICKETTS • *Building Design and Construction Handbook*
NEWMAN • *Design and Construction of Wood Framed Buildings*
NEWMAN • *Standard Handbook of Structural Details for Building Construction*
SHARP • *Behavior and Design of Aluminum Structures*

Building Materials Technology

Structural Performance and Environmental Impact

L. Reed Brantley, Ph.D., Ch.E.

Ruth T. Brantley, M.Ed., C.H.E.

McGraw-Hill, Inc.

New York San Francisco Washington, D.C. Auckland Bogotá
Caracas Lisbon London Madrid Mexico City Milan
Montreal New Delhi San Juan Singapore
Sydney Tokyo Toronto

Library of Congress Cataloging-in-Publication Data

Brantley, L. Reed.
   Building materials technology : structural performance and environmental impact / L. Reed Brantley, Ruth T. Brantley.
      p.   cm.
   Includes index.
   ISBN 0-07-007265-5
   1. Building materials. 2. Building materials—Environmental aspects. I. Brantley, Ruth T. II. Title.
TA403.6.B73   1995
691—dc20   95-11165
                                                                     CIP

Copyright © 1996 by McGraw-Hill, Inc. All rights reserved. Printed in the United States of America. Except as permitted under the United States Copyright Act of 1976, no part of this publication may be reproduced or distributed in any form or by any means, or stored in a data base or retrieval system, without the prior written permission of the publisher.

1 2 3 4 5 6 7 8 9 0   DOC/DOC   9 0 0 9 8 7 6 5

ISBN 0-07-007265-5

The sponsoring editor for this book was Larry S. Hager, the editing supervisor was Stephen M. Smith, and the production supervisor was Suzanne W. B. Rapcavage. It was set in Century Schoolbook by Estelita F. Green of McGraw-Hill's Professional Book Group composition unit.

Printed and bound by R. R. Donnelley & Sons Company.

McGraw-Hill books are available at special quantity discounts to use as premiums and sales promotions, or for use in corporate training programs. For more information, please write to the Director of Special Sales, McGraw-Hill, Inc., 11 West 19th Street, New York, NY 10011. Or contact your local bookstore.

This book is printed on acid-free paper.

Information contained in this work has been obtained by McGraw-Hill, Inc., from sources believed to be reliable. However, neither McGraw-Hill nor its authors guarantee the accuracy or completeness of any information published herein and neither McGraw-Hill nor its authors shall be responsible for any errors, omissions, or damages arising out of use of this information. This work is published with the understanding that McGraw-Hill and its authors are supplying information, but are not attempting to render engineering or other professional services. If such services are required, the assistance of an appropriate professional should be sought.

Contents

Preface xv
Acknowledgments xvii

Chapter 1. Introduction to Science and Technology   1

1.1	Introduction	2
1.2	Spontaneous Environmental Degradation	2
1.3	The Restless Air Molecules Surrounding Us	3
1.4	Basic Science and Technology	4
	1.4.1 Visualizing Models	4
	1.4.2 Model of the Atom	5
	1.4.3 Electron Bonds and Octet Structure	6
	1.4.4 Electron Shells of Common Elements	8
	1.4.5 Periodic Table of the Elements	10
1.5	Inorganic Materials	12
	1.5.1 Nomenclature	13
	1.5.2 Chemistry of Aluminum and Iron	14
	1.5.3 Atomic Weight (Atomic Mass)	15
	1.5.4 The Mole Concept	15
	1.5.5 Chemical Equations and Interpretation	16
	1.5.6 Chemical Equilibrium	16
	1.5.7 Ionic Equilibrium in Water	17
1.6	Organic Materials	18
	1.6.1 The Synthetic World of Organic Chemistry	19
	1.6.2 Scientific Language	19
	1.6.3 The Language of Organic Chemistry	20
	1.6.4 Aliphatic Hydrocarbons with Other Elements	22
	1.6.5 Aromatic Series of Carbon Compounds	23
1.7	Common Synthetic Polymer Types	24
	1.7.1 Thermoplastic Polymers	25
	1.7.2 Thermosetting or Cross-Linked Polymers	27
Terms Introduced		28
Questions and Problems		29
Suggestions for Further Reading		29

Chapter 2. Site Investigation, Preparation, and Environment 31

    2.1 Introduction 31
    2.2 Classification of Soils 32
    2.3 Site Investigation 33
    2.4 Soil Structure, Composition, and Behavior 34
    2.5 Structure of Clay 35
    2.6 Site Preparation for Structural Foundations 37
        2.6.1 Soil Drainage 39
        2.6.2 Compaction Methods 40
        2.6.3 Chemical Treatment of Clay Soils 40
        2.6.4 Piles and Caissons 41
    2.7 Our Environmental Responsibilities 44
        2.7.1 Recycling Used Building Materials 44
        2.7.2 Occupational Safety and Health Administration 45
        2.7.3 Environmental Impact Statement 46
        2.7.4 Environmental Response, Compensation, and Liability 46
        2.7.5 Structural Environmental Hazards 48
    2.8 Soil as a Growth Medium 48
        2.8.1 The Physical Nature of Soils 49
        2.8.2 Permeability of Soil to Water and Nutrients 50
        2.8.3 Major Soil Nutrients 51
        2.8.4 Trace Soil Nutrients 53
    2.9 Hazards of Pesticide Usage 55
    2.10 Irrigation and Industrial Waste Pollution 56
    2.11 Environmental Considerations 57
    Terms Introduced 58
    Questions and Problems 58
    Suggestions for Further Reading 59

Chapter 3. Cement and Concrete 61

    3.1 Introduction 62
    3.2 Cementitious Materials 62
    3.3 Formulas of Cement Compounds 62
    3.4 Portland Cement 63
        3.4.1 Types of Portland Cement 64
    3.5 Hydration, Setting, and Hardening 65
        3.5.1 Setting of Gypsum Plaster 65
        3.5.2 Setting of Lime 66
        3.5.3 Setting and Hardening of Portland Cement 66
    3.6 Concrete 67
        3.6.1 Curing Fresh Concrete 68
        3.6.2 Water/Cement Ratio 68
        3.6.3 Porosity 69
    3.7 Aggregates of Concrete 69
    3.8 Admixtures in Concrete 70
        3.8.1 Water Reducers 70
        3.8.2 Accelerators 72
        3.8.3 Set Retarders 73
        3.8.4 Air-Entrainment Agents 74

3.9	Mineral Admixtures	74
3.10	Reinforced Concrete	74
3.11	Prestressed Concrete	78
3.12	Polymer Concrete	78
3.13	Composites	79
3.14	Environmental Durability	80
	3.14.1 The Corrosive Effect of Salts	81
	3.14.2 Sulfating	82
	3.14.3 Other Destructive Gases	82
3.15	Corrosion of Steel Reinforcement	83
	3.15.1 Passive Steel	83
	3.15.2 Multiple Anodes and Cathodes	83
	3.15.3 Rusting of Rebars in Concrete	84
	3.15.4 Chloride-Induced Corrosion	84
	3.15.5 Corrosion Protection	85
3.16	Building Problems Related to Concrete	85
3.17	Health Hazards	87
	Terms Introduced	88
	Questions and Problems	88
	Suggestions for Further Reading	89

Chapter 4. Masonry   91

4.1	Introduction	92
4.2	Early Masonry	92
4.3	Modern Masonry	93
4.4	Masonry Mortar	95
4.5	Ingredients of Mortar	97
	4.5.1 Portland Cement Used in Mortar	97
	4.5.2 Blended Cement	97
	4.5.3 Masonry Cements	98
	4.5.4 Lime	98
	4.5.5 Aggregates	98
	4.5.6 Water	99
	4.5.7 Admixtures	99
4.6	Physical Properties of Mortar	99
4.7	Thickness for Optimum Bond Strength	100
4.8	Grout Masonry	100
4.9	Concrete Structures	101
4.10	Masonry Units	101
4.11	Concrete Blocks	102
	4.11.1 Classification	102
	4.11.2 Manufacture	103
	4.11.3 Aggregates	103
	4.11.4 Admixtures	103
	4.11.5 Properties	104
4.12	Cement Bricks	104
4.13	Clay Masonry Units	104
	4.13.1 Classification of Clay Units	104
	4.13.2 Clay Bricks	105

	4.13.3	Clay Tile	106
	4.13.4	Terra Cotta Tile	107
4.14	Stone Masonry	107
4.15	Glass Blocks	111
4.16	Plaster	112
4.17	Restoring Masonry Surfaces	112
	4.17.1	Water Cleaning	113
	4.17.2	Abrasive Cleaning	115
	4.17.3	Chemical Cleaning Agents	115
4.18	Hazardous Chemicals	116
4.19	OSHA Responsibilities for Workers	117
4.20	Masonry Problems and Their Prevention	117
4.21	Environmental Hazards	118
Terms Introduced	118
Questions and Problems	118
Suggestions for Further Reading	119

Chapter 5. Metals   121

5.1	Introduction	122
5.2	Metals in Construction	123
5.3	Metallic Properties, Bonding, and Structure	123
5.4	Metallic Structures	124
	5.4.1 Examples of Metallic Structures	124
	5.4.2 Examples of HCP Crystal Structure	125
	5.4.3 Examples of FCC Crystal Structure	127
	5.4.4 Example of BCC Crystal Structure	129
5.5	Solid-Solution Metallic Structures	129
5.6	Formation of Solid Solutions	130
5.7	Ferrous Metal Structures	131
5.8	Basic Phase Diagrams	132
	5.8.1 Simple Eutectic Formation	132
	5.8.2 Compound Formation	133
	5.8.3 Solid-Solution Formation	133
5.9	Allotropic Forms of Iron	133
5.10	Smelting of Iron Ores	134
5.11	Iron-Carbon Phase Diagram	134
5.12	Iron and Its Alloys	136
	5.12.1 Austenite Formation Temperature	137
	5.12.2 Ferrite Formation Temperature	137
	5.12.3 Cementite Formation Temperature	137
	5.12.4 Pearlite Formation Temperature	138
	5.12.5 Martensite Formation Temperature	138
	5.12.6 Bainite Formation Temperature	138
	5.12.7 Stainless-Steel Alloys	139
5.13	Work Hardening	141
5.14	Heat Treatment	141
5.15	Annealing	142
	5.15.1 Quenching	142
	5.15.2 Tempering	142

5.16	Common Nonferrous Alloys	143
	5.16.1 Aluminum-Silicon Alloys	144
	5.16.2 Aluminum-Copper Alloys	144
	5.16.3 Aluminum-Magnesium Alloys	144
	5.16.4 Copper and Its Alloys	145
	5.16.5 Copper-Zinc Alloys	145
	5.16.6 Copper-Tin Alloys	145
	5.16.7 Copper-Nickel Alloys	146
	5.16.8 Lead-Tin Alloys	146
	5.16.9 High-Performance Metal Composites	146
5.17	Metallic Corrosion	147
5.18	The Corrosion Process	148
5.19	Example of the Oxidation of Aluminum	148
5.20	Galvanic Corrosion between Dissimilar Metals	149
	5.20.1 Harnessing Corrosion in a Battery	149
	5.20.2 Example of Galvanic Corrosion	150
	5.20.3 Example of a Sacrificial Anode	151
5.21	Standard Electrode Potential Series	151
5.22	Local or Nonuniform Corrosion	153
5.23	Corrosion-Resistant Metals and Alloy Steels	155
5.24	Don't Give Up	157
5.25	Balancing Redox Equations	157
5.26	Displacement Series	158
5.27	Prediction of Galvanic Corrosion	158
	5.27.1 Example of a Concentration Cell	159
	5.27.2 Sacrificial Anode Cell	160
5.28	Surface-Corrosion Protection	160
5.29	Corrosion Protection from Circulating Water	161
5.30	Metal Cladding	161
5.31	Electrodeposition of a Metallic Coating	162
5.32	Galvanic Protection with Sacrificial Anodes	163
5.33	Example of a Zinc Sacrificial Anode	163
5.34	Sacrificial Metallic Pigment Coating	165
5.35	Corrosion Protection with Applied Voltage	165
5.36	Passivation by Anodizing	166
5.37	Organic Protective Prime Coat	166
5.38	Phosphoric Acid Treatment of Metallic Surfaces	167
5.39	Corrosion-Protective Pigments	167
5.40	Chemical Conversion Coatings	168
5.41	Building Problems Related to Metals	169
	Terms Introduced	169
	Questions and Problems	170
	Suggestions for Further Reading	171

Chapter 6. Wood, Polymers, Adhesives, and Plastics — 173

6.1	Introduction	174
6.2	Classification of Woods	175
6.3	Composition of Wood	175

	6.3.1 Cellulose	175
	6.3.2 Pentosan and Lignin	176
6.4	Structural Panel Composites	176
	6.4.1 Plywood	176
	6.4.2 Particleboard	177
6.5	Bacterial Destruction of Wood	178
6.6	Fungal Destruction of Wood	178
	6.6.1 White Rot	179
	6.6.2 Brown Rot	179
	6.6.3 Soft Rot	179
	6.6.4 Rot Protection	179
6.7	Deterioration by Insects	180
	6.7.1 Wood Inhabitants	180
	6.7.2 Carpenter Ants	181
	6.7.3 Carpenter Bees	181
	6.7.4 Termites	182
6.8	Prevention and Control of Termites	184
	6.8.1 Basaltic Termite Barrier	185
	6.8.2 Cell Treatment of Wood	185
	6.8.3 Protection by Fumigation	186
6.9	Paper Technology	186
	6.9.1 Modern Papermaking	186
	6.9.2 Self-Destruction of Modern Paper	187
	6.9.3 Treatment of Acid Paper	187
6.10	Polymers	187
	6.10.1 Natural and Synthetic Polymers	188
	6.10.2 Structure of Linear Polymeric Chains	188
6.11	Physical Properties of Polymers	190
	6.11.1 Permeability of Polymers	190
	6.11.2 Coefficient of Expansion	192
	6.11.3 Glass-Transition Temperature	192
	6.11.4 Lowering the Glass-Transition Temperature	192
6.12	Example of Glass-Transition Temperature	193
6.13	Solvation of Linear Polymers	193
6.14	Plasticization of Cross-Linked Polymers	193
6.15	Elastomers, Fibers, and Plastics	194
	6.15.1 Elastomers	194
	6.15.2 Fibers	194
	6.15.3 Plastics	196
6.16	Example of Polymer Physical Properties	197
6.17	Plastics	198
	6.17.1 Flat Sheet Plastic	199
	6.17.2 Availability and Capability	200
	6.17.3 Limiting Physical Properties	200
	6.17.4 Residential Plastic Materials	201
6.18	Natural and Synthetic Rubber	202
6.19	Adhesives	202
	6.19.1 Cohesive Strength of Adhesives	203
	6.19.2 Adherence of Adhesives	203
	6.19.3 Fluidity of Liquid Adhesives	204
	6.19.4 Wettability	204

6.20	Types of Adhesives	205
	6.20.1 Organic Solvent-Thinned Adhesives	205
	6.20.2 Latex Adhesives (Water Emulsions)	205
	6.20.3 Water-Dispersed Adhesives	206
	6.20.4 Two-Package Adhesives	206
6.21	Advances in Structural Adhesives	206
	6.21.1 Rubber-Modified Resins	207
	6.21.2 Acrylic Adhesives	207
	6.21.3 Particulate-Reinforced Resins	210
	6.21.4 Silane Surface Primer	210
6.22	Adhesives for Composites	210
	6.22.1 Cross-Linked Adhesives	210
	6.22.2 Novalak Resins	211
	6.22.3 The Resol, Resitol, and Resital Stages	211
6.23	Health Hazards	211
6.24	Recycling of Wood, Paper, and Plastics	212
	Terms Introduced	214
	Questions and Problems	215
	Suggestions for Further Reading	215

Chapter 7. Roofing, Sealants, and Fire Protection — 217

7.1	Introduction	218
7.2	Roofing	218
7.3	Understanding Emulsions	221
	7.3.1 Example of Formation of a Colloidal Solution	223
	7.3.2 Example of Formation of an O/W Emulsion	223
	7.3.3 Example of Cracking an Emulsion	224
7.4	Membrane Materials	224
	7.4.1 Asphalt Derivatives	224
	7.4.2 Asphalt Emulsions	225
	7.4.3 Asphaltic Bitumens	225
	7.4.4 Blown Asphalt	225
	7.4.5 Cracked Asphalt	225
	7.4.6 Cut-Back Asphalt	225
7.5	Built-Up Membranes	226
	7.5.1 Single-Ply Membranes	226
	7.5.2 Foamed Plastic Membranes	227
7.6	Water-Shedding Systems	227
7.7	Residential Roofs	228
	7.7.1 Shakes and Shingle Roofs	228
	7.7.2 Asphalt Shingles	229
	7.7.3 Grades of Asphalt Roofing Materials	229
	7.7.4 Wood Shingles	229
	7.7.5 Mineral-Fiber Shingles	230
7.8	Insulation of Roofs	230
7.9	Typical Problems and Correction	231
7.10	Government Restrictions	233
7.11	Recycling Roofing Materials	234
7.12	Warranties	234

7.13	Sealants	235
	7.13.1 Movement Capability	236
	7.13.2 Hardness	237
	7.13.3 Adhesive Promoters	237
	7.13.4 Primers	237
7.14	Sealant Types	238
	7.14.1 Oil-Based Caulks	238
	7.14.2 Polysulfide Sealants	239
	7.14.3 Butyl Sealants	239
	7.14.4 Acrylic Sealants	240
	7.14.5 Urethane Sealants	241
	7.14.6 Silicone Sealants	242
7.15	Fire Protection	243
	7.15.1 Materials at High Temperatures	243
	7.15.2 Wood Exposed to Fire	244
	7.15.3 Concrete Exposed to Fire	245
	7.15.4 Steel Exposed to Fire	245
	7.15.5 Polysteel Forms	246
7.16	Fire Security	246
	7.16.1 Fire Detection	247
	7.16.2 Containment of Fire	247
	7.16.3 Protection of Structural Elements	248
7.17	Repair of Damage	248
7.18	Health Hazards	250
	Terms Introduced	251
	Questions and Problems	251
	Suggestions for Further Reading	251

Chapter 8. Glass   253

8.1	Introduction	254
8.2	Glass Formation	254
8.3	Properties of Glass	255
	8.3.1 Viscosity	255
	8.3.2 Glass-Transition Temperature	256
	8.3.3 Nucleation and Crystallization	256
	8.3.4 Tensile Strength	257
8.4	Basic Silica Structures	257
	8.4.1 Amorphous Solids	258
	8.4.2 Crystalline Silica	258
	8.4.3 Silica-Type Structures	258
8.5	Kinds of Glass	259
	8.5.1 Soda-Lime Silica (Soft) Glass	259
	8.5.2 Plate Glass	260
	8.5.3 Borosilicate Glass	260
	8.5.4 Lead Glass	260
	8.5.5 Optical Glass	260
	8.5.6 Photosensitive Glass	261
8.6	Safety Glass	261
	8.6.1 Wired Glass	261
	8.6.2 Tempered Glass	261

	8.6.3 Chemically Toughened Glass	264
	8.6.4 Laminated Glass	264
	8.6.5 Special-Purpose Laminates	265
8.7	Insulated Glass	265
8.8	Solar-Control Glass	267
	8.8.1 Tinted Glass	267
	8.8.2 Coated Glass	269
8.9	Structural Glazing	269
8.10	Glass Fibers	270
8.11	Glass Ceramics	271
8.12	Recycling	273
Terms Introduced		273
Questions and Problems		273
Suggestions for Further Reading		274

Chapter 9. Protective and Decorative Finishes   275

9.1	Introduction	276
9.2	Film Formers	276
	9.2.1 Alkyd Resins	277
	9.2.2 Vinyl Resins	278
	9.2.3 Polyurethane Resins	278
	9.2.4 Acrylic Resins	279
	9.2.5 Epoxy Resins	280
	9.2.6 Silicone Resins	281
9.3	Paint Solvents	281
9.4	Pigments	282
9.5	Additives	282
9.6	Surface Preparation	283
9.7	Organic Coating Systems	285
	9.7.1 Selecting the Right Paint	286
	9.7.2 Prime Coat	287
	9.7.3 Water-Emulsified Paints	287
	9.7.4 Typical Paint System	288
9.8	Physical Properties of Paints	288
	9.8.1 Solubility of Organic Substances	289
	9.8.2 Solubility Parameters of Liquids	289
	9.8.3 Solubility Parameters of Organic Polymers	289
	9.8.4 Polar and Nonpolar Substances	290
	9.8.5 Viscosity of Polymer Solutions	290
	9.8.6 Rheology, Thixotropy, and Dilatancy	291
	9.8.7 Gloss	291
	9.8.8 Transition Temperature	292
	9.8.9 Adhesion	292
9.9	Paint Problems and Correction	293
9.10	Lead Toxicity and Its Abatement	295
9.11	Health Hazards	296
Terms Introduced		296
Questions and Problems		297
Suggestions for Further Reading		297

Chapter 10. Design, Interiors, Furnishings, and Equipment — 299

- 10.1 Introduction — 299
- 10.2 Computer-Aided Design — 300
- 10.3 Indoor Air Quality — 300
- 10.4 Lighting Design — 302
- 10.5 Office Equipment and Home Appliances — 303
    - 10.5.1 Office Machines and Supplies — 303
    - 10.5.2 Refrigerators and Ozone — 303
    - 10.5.3 Saving Energy in the Home — 303
    - 10.5.4 Home Heating Appliances — 304
    - 10.5.5 Home Security — 305
- 10.6 Upholstery and Drapery Fabrics — 308
- 10.7 Wallcoverings — 309
- 10.8 Furniture — 310
- 10.9 Flooring — 310
- 10.10 Carpets — 312
- 10.11 Cleaning Agents — 312
- 10.12 Outdoor Decks — 313
- 10.13 What about Metrics? — 313
- Terms Introduced — 313
- Questions and Problems — 314
- Suggestions for Further Reading — 315

Appendix A 317
Appendix B 321
Index 323

Preface

Anyone in the building industry or a related field, remodeling, repairing, or maintaining buildings most likely is overwhelmed by the rapid development of new building materials and the improvement of existing materials. To evaluate the claims made about a material and to select the best material for the intended purpose, it is helpful to become familiar with the composition and structure of each material that determines its properties and performance. This book provides information about the best use of a wide variety of modern building materials.

This book is specially designed for the needs of student and professional architects, specifications writers, home economists, interior designers, industrial designers, landscape architects, industrial arts teachers, sales personnel, other motivated professionals, and homeowners who have a limited background in science and need information on specific items. Those familiar with the Construction Specifications Institute (CSI) format will find the arrangement of topics convenient.

A unique feature of this book is its combining of the technology of building materials with the essential concepts of physical science and engineering, together with health hazards and impact on the environment. After an introductory review and update of the scientific concepts needed for understanding, each chapter contains the science and technology appropriate to a particular material. Some background in physical science is useful, but is not essential.

Experienced professionals will find this blending of the basic science and technology of building materials to be refreshing in its application of scientific principles to practical experience. Maintenance-oriented readers will find the listing of common problems and their solutions to be quite useful. Suggestions for further reading, at the end of each chapter, provide an opportunity to pursue a topic in depth.

Familiarity with basic concepts is important when evaluating the claims made about and the limitations of new products. The information found in this book will help the reader avoid being open to litigation over the accidental failure or health hazards of a product and prepare him or her for the flood of synthetic building materials reaching the market now and into the next century.

L. Reed Brantley
Ruth T. Brantley

Acknowledgments

The authors dedicate this book to Elmer E. Botsai, FAIA, Professor and former Dean of the School of Architecture, The University of Hawaii at Manoa, who had the foresight to require in the undergraduate architecture curriculum a basic course in the chemistry of building materials. His continued encouragement made this book possible.

Special recognition is due to the following for suggestions relating to their fields of expertise: Dr. Frank L. Lambert, Professor Emeritus of Organic Chemistry, Occidental College; Gale Farquhar, Corrosion Engineer, Texaco, Inc.; Kenneth N. Edwards, Principal, Dunn-Edwards Corporation, Inc.; Dr. Alvin R. Zigman, Director of Technical Computations, 3M; and Dr. Glenn T. Seaborg, Nobel Laureate, Lawrence Berkeley Laboratory, University of California.

For assisting with their advice, endless patience, and encouragement, our sincere appreciation is due to Prof. William G. Messer, Prof. Frederick L. Creager, Prof. Mary Ellen Des Jarlais, Prof. Arthur Mori, Philip G. Heyenga, Helen S. Norton, Christopher Messer, Jo Miller, Darla Rogge, Randy Y. Tanaka, and Dennis Swart.

Building Materials Technology

Chapter 1

Introduction to Science and Technology

1.1 Introduction	2
1.2 Spontaneous Environmental Degradation	2
1.3 The Restless Air Molecules Surrounding Us	3
1.4 Basic Science and Technology	4
1.4.1 Visualizing Models	4
1.4.2 Model of the Atom	5
1.4.3 Electron Bonds and Octet Structure	6
1.4.4 Electron Shells of Common Elements	8
1.4.5 Periodic Table of the Elements	10
1.5 Inorganic Materials	12
1.5.1 Nomenclature	13
1.5.2 Chemistry of Aluminum and Iron	14
1.5.3 Atomic Weight (Atomic Mass)	15
1.5.4 The Mole Concept	15
1.5.5 Chemical Equations and Interpretation	16
1.5.6 Chemical Equilibrium	16
1.5.7 Ionic Equilibrium in Water	17
1.6 Organic Materials	18
1.6.1 The Synthetic World of Organic Chemistry	19
1.6.2 Scientific Language	19
1.6.3 The Language of Organic Chemistry	20
1.6.4 Aliphatic Hydrocarbons with Other Elements	22
1.6.5 Aromatic Series of Carbon Compounds	23
1.7 Common Synthetic Polymer Types	24
1.7.1 Thermoplastic Polymers	25
1.7.2 Thermosetting or Cross-Linked Polymers	27
Terms Introduced	28
Questions and Problems	29
Suggestions for Further Reading	29

1.1 Introduction

What unique properties distinguish one building material from another? What inner structure gives this material its special properties and performance, even in hostile environments? How do building materials and their uses pose a threat to construction personnel, the eventual occupants, and the environment? The purpose of this book is to find the answers to these and many other questions.

In addition to common building materials used every day, modern technology has produced a bewildering array of new materials, such as new developments in cement composites, metal alloy composites, corrosion protection, glass ceramics, protective coatings, and plastics. Using these new materials without having a scientific background to fully appreciate their limitations and applications has led to tragic building failures, injuries, and expensive lawsuits. To provide background, an introduction to the relevant principles of science and technology is given in this chapter. Specific information relating to each class of building material is provided in later chapters.

Using guidelines provided by the Occupational Safety and Health Administration (OSHA), contractors and homeowners are increasing their efforts to reduce accidents and health disorders caused by exposure to hazardous chemicals in the materials they use. On a larger scale, the Environmental Protection Agency (EPA) concentrates on the potential hazards of chemicals from building materials that can have long-term harmful effects on construction personnel, the eventual occupants of the structures, and the environment. Guidelines of both the American Society for Testing and Materials (ASTM) and the Universal Building Code (UBC) are frequently incorporated in building codes. All of these agencies are included in later discussions of the individual common building materials and in accord with our realization of social responsibility for our planet. An overview of these efforts is presented with each building material.

At the end of each chapter you will find a list of the technical terms that have been introduced. The Index provides a reference for each of these terms. Also, there are questions and problems to test your understanding of the material and suggestions for further reading if you wish to study a topic in depth.

1.2 Spontaneous Environmental Degradation

Why do human-made materials and structures gradually corrode and disintegrate into the ores and chemicals from which they originated?

Steel rusts, cement and stone crack and crumble, and coatings deteriorate. However, universal degradation, a law of nature, cannot be halted or reversed. Some of this degradation, as described in later chapters, can be delayed by better maintenance or choice of appropriate materials, or both.

This degradation experience is summarized in two laws of thermodynamics. The first law states that mass and energy cannot be created or destroyed. Conversion from one to the other can occur only under special conditions found by exposure to the sun and in nuclear reactors. The second law states that spontaneous corrosion and disintegration of materials is a universal law of nature and cannot be halted or reversed. If we are to delay this natural process, our challenge is to consider the chemical and physical properties of the materials required for each job.

Where are these natural, herculean forces of nature that deteriorate our structural materials? To illustrate their presence, let us first consider the invisible molecules in the air surrounding us.

1.3 The Restless Air Molecules Surrounding Us

You are aware of the things you can see, smell, and touch, but what about the air you breathe? Can you visualize air as a mixture of submicroscopic molecules bouncing off your body at a speed of about 1000 miles per hour?

By volume, air is mostly three parts nitrogen and one part oxygen and contains traces of other substances, such as carbon dioxide and water vapor. Each is essential to some form of life. Carbon dioxide, vital to plant life and responsible for the global greenhouse effect, acts as a giant blanket regulating the amount of heat emanating from the sun and retained by our planet.

The oxygen molecule has two oxygen atoms bonded together as a unit. This molecule has a diameter of about one hundred millionth of an inch (254 millionths of a centimeter). With a mass so small that 1.8 trillion molecules weigh one trillionth of an ounce (28 trillionths of a gram) it is little wonder that we are unaware of their presence. Nitrogen molecules are slightly smaller. Yet the combined efforts of these invisible lilliputian hit-and-run air molecules exert a pressure at sea level of 14.7 pounds per square inch (149 pascals per 2.5 square centimeters), pressing on each square inch of our bodies.

One molecule of a room-temperature gas moving at a velocity of nearly 1 mile (1.6 kilometers) per second travels only a distance of 200 times its diameter before it bumps into another molecule. Some of these collisions slow the molecule; others speed it up. Although the majority of the molecules has an average velocity determined by temperature, individual molecules differ greatly from one another in their velocity and the energy represented. Therefore a tiny fraction of them has a large velocity corresponding to a much higher temperature.

In general, molecular velocities in a gas increase with an increase in temperature. It follows that the frequency of collision and the thermal energy increase as well. Since reactive atoms of a molecule must come into contact in order for them to react, the number of molecules that can chemically react will increase as the temperature increases.

Individual molecules in a solid or a liquid have thermal energy as do molecules of a gas, except their motion is much more restricted. The molecules in building materials differ from one another in their internal energy. Being tightly packed together, their motion is limited to energetic oscillation, rotation, and bond vibration.

How does this relate to our apparently quiescent building materials, which are composed of restless atoms and molecules? We are familiar with dangers of collapse from an overloaded structure, or sparks or flames around flammable liquids, or mixtures of explosive gases. Our apparently stable building structures are vulnerable in a different way. There are always a few exceptional molecules with very high energies ready to react at room temperature. These molecules need access to moisture, salt, oxygen, and sunlight from the environment. For example, iron rusts and disintegrates slowly and relentlessly.

1.4 Basic Science and Technology

Chemical bonds hold atoms and molecules together in a variety of structures. The atoms may be arranged in amorphous, crystalline, or metallic structures, or in polymeric chains. How do these chemical bonds produce these structures with their unique properties and performances? The following basic ideas of science and technology will help you to find the answers to these questions.

1.4.1 Visualizing models

Visualize the atoms in a metal beam. As the beam is flexed, the molecules glide over one another; not all of the molecules return to their original places. This could cause the beam to fracture or develop an area susceptible to corrosion. From this model of a solid, just imagine how these confined, energetic little atoms can affect the performance of structures.

You are challenged to use your imagination to build and use models of chemical materials and their structures. Consider an easy example. How does a model of restless molecules predict how water escapes from wet clothes as they hang on a clothes line? How can water evaporate when it is so far from its boiling point of 100°C (212°F)? How can clothes dry when hanging outside in freezing weather?

Can you picture a liquid water molecule that is bumped so many times in the same direction that it moves much faster than the average? If this picture does not appear to be useful, pretend we are talking about just-placed concrete in which too much water has been added. A few of the water molecules will have a velocity high enough to escape when they reach the surface. They must also overcome the attraction of their fellow molecules in order to soar off into the air as water vapor. Would a higher ambient temperature mean higher average velocities and faster evaporation? Had you thought of the forces holding these atoms together in the molecule? And what does the structure of these atoms look like? For answers to these questions, let us construct a simple model of an atom.

1.4.2 Model of the atom

An element is usually described as a substance that cannot be further subdivided into different substances by ordinary chemical means. An atom is the simplest, smallest subdivision of an element that still has the chemical properties of the element.

A working model of the atom needs only three fundamental particles: protons, neutrons, and electrons. Like our solar system with planets circling the sun, the model of the atom has electrons orbiting around the nucleus. Like the sun in our solar system, most of the mass of the atom is in this nucleus. The nucleus is composed of protons and neutrons. These two nuclear particles have practically the same mass, but the proton has an additional unit of positive charge. The number of protons (units of positive electrical charge) in the nucleus is called the atomic number of the element. Each element has a unique atomic number and is identified by this number. The atomic mass of the element is approximately equal to the sum of the masses of its protons and neutrons because the electrons have so little mass.

When the atoms of a given element have different numbers of neutrons, the atoms are called isotopes of the element. For example, the element chlorine has two common isotopes. Both isotopes have 17 nuclear protons and an atomic number of 17 that identifies them as chlorine atoms. One isotope has 18 neutrons in its nucleus and the other isotope has 20 neutrons in its nucleus. Thus one of the isotopes has a mass of 35 atomic mass units, whereas the other has 37 atomic mass units. The isotope with 17 neutrons is more abundant. Therefore, the average atomic mass of chlorine is 35.47 atomic mass units.

The electron is the lightest of the three atomic particles with about $1/1846$ of the mass of a proton, but it has an opposite (negative) unit of charge. An atom has an equal number of electrons and protons, making it electrically neutral. This means that the number of orbiting electrons must be equal to the number of protons in the nucleus of the atom. The number of protons in the nucleus, or the total number of outer electrons of an atom, is equal to its atomic number.

In our simplified model of the atom, the outer electrons are held in their orbits around the nucleus by the attraction between the negative charge of the electron and the positive charge of the nucleus. Electrons have the least energy and are more stable in the orbits closest to the nucleus. Accordingly, the orbits normally filled in first by electrons are those nearest the nucleus.

The path that each family of electrons sweeps out as they orbit around the nucleus makes up a layer called an electron shell. Within a shell the first two paths are circular orbits, whereas the rest are elliptical. The electron shells are labeled alphabetically, starting with the letter K. The innermost layer, or first shell of electrons, is the K shell. (This K shell is said to be named after Katrina, the daughter of the physicist Niels Bohr.) The K shell can hold only two electrons. The next larger diameter electron shell continues alphabetically with the letter L. The L shell has room for eight electrons. The third, or M, shell can hold 18 electrons, and the N shell has places for 32 electrons. The eight innermost electrons in each electron shell are chemically stable and do not usually enter into reactions. (Only two electrons are needed for hydrogen and helium.) The electrons in the outer unfilled shell enter into reactions and determine the chemical reactivity of the element. This model is a modified form of the Bohr theory of the atom.

1.4.3 Electron bonds and octet structure

The outer, or valence, shell contains the electrons that bond atoms together to form compounds. The usual electron bond consists of a pair of electrons. These two electrons should repel one another because of their negative charges. But the electrons in the bond spin about their axes to form magnets which attract one another. The magnetic attraction is stronger than the repulsion of their negative

Introduction to Science and Technology   7

charges. Thus an electron pair can act as an imaginary electronic glue to bond atoms together.

The number of electrons in the valence shell determines the chemical behavior of the element. For example, the most unreactive elements are the noble gases: helium, neon, argon, and krypton. Helium is so small it becomes inert with its two outer K electrons. The other inert gases have an additional six electrons to make a total of eight electrons. This is the basis for the octet structure.

For practical purposes, four pairs of electrons in the valence shell, called an octet, cause the atom to become chemically unreactive. The octet structure has shown that a reactive atom can become chemically unreactive by either gaining or losing electrons to form an octet of electrons in its outer valence shell.

Because of the inertness of the inner electron shells beneath the completed octet, it is customary to include them in the symbol of the element when writing electron dot formulas. Only valence electrons available to form compounds need to be indicated as dots. Examples are Ca: for the calcium atom and Na• for the sodium atom.

Electrons form two important kinds of chemical bonds. In one kind, valence electrons are shared with another atom to form a covalent or shared electron bond. For example, Fig. 1.1 shows the carbon atom with four valence electrons and the chlorine atom with seven. Both atoms must form an octet of eight electrons around them to make a stable, unreactive electron shell. Each carbon atom has four electrons to share. Each chlorine atom needs one more electron. One carbon atom shares its electrons with four chlorine atoms to form carbon tetrachloride (CCl_4).

An example in Fig. 1.1 is H_2O, the water molecule. It has two shared electron bonds. To illustrate, start with the atoms. The oxygen atom has six valence electrons, so oxygen needs to gain two more electrons. Oxygen shares an electron from each of the two hydrogen atoms. In this way oxygen completes its octet and hydrogen gets its electron pair.

Another common kind of bond is the ionic bond. It is formed when one atom is stripped of its valence electrons to complete the octet of

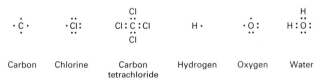

Figure 1.1   Model of covalent compounds.

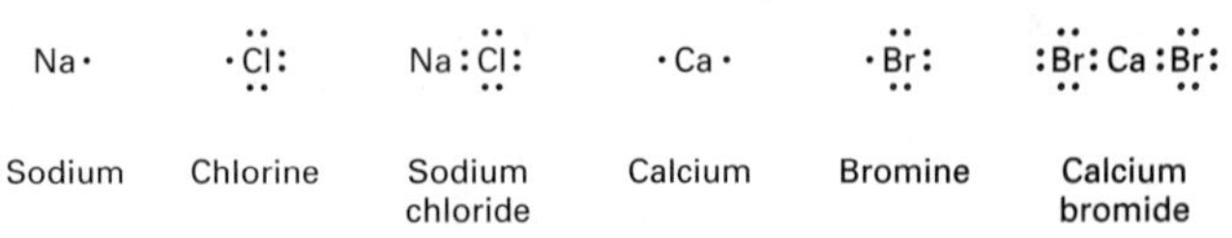

Na·   ·C̈l̈:   Na:C̈l̈:   ·Ca·   ·B̈r̈:   :B̈r̈:Ca:B̈r̈:

Sodium   Chlorine   Sodium chloride   Calcium   Bromine   Calcium bromide

Figure 1.2 Model of ionic compounds.

the other more electron-thirsty atom. This is illustrated in Fig. 1.2 by the reaction of sodium and chlorine to form sodium chloride, or table salt. The atoms of a metallic element, such as sodium, have only one electron in their valence shell whereas chlorine has seven. Usually a metallic atom that has only one or two electrons in its valence shell will lose them to a nonmetallic atom that lacks one or two electrons to complete its octet.

Sodium readily gives up its single valence electron to expose its stable inner shell of eight electrons. This allows the nonmetallic chlorine atom to complete its octet. In the process the sodium atom loses one negative electron to the chlorine atom. This gives sodium a positive charge and chlorine a negative charge.

Charged atoms, such as sodium and chlorine, are called ions. The attraction between these oppositely charged ions is the ionic bond that holds sodium chloride together.

The calcium bromide molecule is another illustration of an ionic bond. Bromine, a nonmetal, is like chlorine in its electron thirst since it also has seven valence electrons. The metallic calcium atom has two valence electrons. Using our simplified model, calcium gives an electron to each of two bromine atoms. By the loss of its two valence electrons, calcium is stripped to the next innermost shell of an octet of eight electrons. The two bromine atoms have each added an electron to complete their octets. The chemical formula is $CaBr_2$. Ionic bonded compounds separate into their charged ions when dissolved in water. For example, the $CaBr_2$ molecule separates into Ca^{++} and two Br^- ions.

The octet structure provides a useful model for many inorganic compounds, but where can you find the number of valence electrons contained in an element? Also, how many different elements exist to form compounds? Table 1.1 will help you answer these questions.

1.4.4 Electron shells of common elements

Table 1.1 contains a simplified arrangement of the common elements, each with its abbreviated chemical symbol, atomic number, atomic mass, and electron shells. (A more complete table of the elements is

Introduction to Science and Technology

TABLE 1.1 Atomic Number, Mass, and Electron Shells*

Element	Symbol	Atomic number	Atomic mass	Electron shell K	L	M	N
Hydrogen	H	1	1.008	1			
Helium	He	2	4.003	2			
Lithium	Li	3	6.941	2	1		
Beryllium	Be	4	9.012	2	2		
Boron	Bo	5	10.81	2	3		
Carbon	C	6	12.01	2	4		
Nitrogen	N	7	14.01	2	5		
Oxygen	O	8	16.00	2	6		
Fluorine	F	9	19.00	2	7		
Neon	Ne	10	20.18	2	8		
Sodium	Na	11	22.99	2	8	1	
Magnesium	Mg	12	24.31	2	8	2	
Aluminum	Al	13	26.98	2	8	3	
Silicon	Si	14	28.09	2	8	4	
Phosphorus	P	15	30.97	2	8	5	
Sulfur	S	16	32.06	2	8	6	
Chlorine	Cl	17	35.45	2	8	7	
Argon	A	18	39.95	2	8	8	
Potassium	K	19	39.10	2	8	8	1
Calcium	Ca	20	40.08	2	8	8	2
Scandium	Sc	21	44.96	2	8	9	2
Titanium	Ti	22	47.90	2	8	10	2
Vanadium	V	23	50.94	2	8	11	2
Chromium	Cr	24	52.00	2	8	13	1
Manganese	Mn	25	54.94	2	8	13	2
Iron	Fe	26	55.85	2	8	14	2
Cobalt	Co	27	58.93	2	8	15	2
Nickel	Ni	28	58.70	2	8	16	2
Copper	Cu	29	63.55	2	8	18	1
Zinc	Zn	30	65.38	2	8	18	2

*For a complete listing of the elements, see App. A.

given in the Appendix.) The electron shells show the electron arrangement of each shell of the atoms. Of special interest is the number of electrons in the outer, unfilled electron shell. These are the valence electrons. The number of electrons in this valence shell is used to predict the chemical nature of the element.

An atom with the atomic number of 1 has the chemical properties of hydrogen. Hydrogen has additional isotopes: deuterium and tritium. Deuterium is found in heavy water. Tritium is the radioactive isotope used in fusion. Helium, with 2 for its atomic number, has two electrons to make a filled K shell. Helium is one of the chemically unreactive noble gases. Lithium, with three protons in its nucleus, has a filled K shell of electrons and an additional reactive electron in its L valence shell. This electron is so unstable that lithium reacts vigorously with water. The hydrogen formed in the reaction spontaneously ignites from the heat liberated.

Beryllium, with two electrons in the L shell, is a less reactive metal than lithium. Boron, with three valence electrons in the L shell, is more of a nonmetal than a metal. Carbon, with four valence electrons in the L shell, is the nonmetal responsible for the vast number of organic compounds made from petroleum, including gasolines and plastics. Nitrogen, with an atomic number of 7, has a filled K shell and five valence electrons for the L shell. As an example, at the high temperature of an automobile engine, nitrogen reacts with oxygen from the air to make the nitrogen oxides that go back into the air to produce smog.

Oxygen, with one more valence electron to make a total of six L electrons, has definite nonmetallic properties. Fluorine, so reactive that it reacts violently with most substances with which it comes into contact, has seven electrons in its L valence shell and needs one more from another atom to complete its octet. Neon, with a total of 10 electrons, has a complete octet of outer electrons in the L shell and is a noble gas. As the process is repeated, the atomic number increases.

Usually only the valence electrons in the outer shell enter into chemical reactions to form compounds. The group of elements starting with potassium, atomic number 19, has electronic orbits with such eccentricity and lower energy that electrons in the next out N electron shell are slightly more stable and start filling the N shell before the M shell is completely filled. This slight difference in energy between electrons in adjacent shells allows such elements as iron (atomic number 26) to have a choice of the number of electrons with which to form compounds, depending on the abundance of the other element. Thus iron rusts in the presence of oxygen and moisture to form either ferrous oxide or ferric oxide. Ferric oxide is the more stable form of iron in the presence of the amount of oxygen found in the air.

Many elements with large atomic numbers have unstable nuclei and are radioactive. Some are so unstable they are not found in nature and have only been made with high-energy particle accelerators. Because they exist for just a fraction of a second, it becomes increasingly more difficult to extend the periodic table of elements any further.

1.4.5 Periodic table of the elements

Years ago the first periodic table had the elements that were known to exist at that time arranged in order of increasing atomic weight. In

TABLE 1.2 Partial Periodic Table of the Elements

Group								Group							
1A	2A	3A	4A	5A	6A	7A	8A	1B	2B	3B	4B	5B	6B	7B	8B
H															He
Li	Be									B	C	N	O	F	Ne
Na	Mg									Al	Si	P	S	Cl	Ar
K	Ca	Sc	Ti	V	Cr	Mn	Fe Co Ni	Cu	Zn	Ga	Ge	As	Se	Br	Kr
Rb	Sr	Y	Zr	Nb	Mo	Tc	Ru Rh Pd	Ag	Cd	In	Sn	Sb	Te	I	Xe
Cs	Ba	La	Hf	Ta	W	Re	Os Ir Pt	Au	Hg	Ti	Pb	Bi	Po	At	Rn
Fr	Ra	Ac													

designing this periodic table, Mendeleyev, a Russian scientist, was probably influenced by the octave sequence of the musical scale. Not only did the periodic table correlate the elements by their chemical properties, but it predicted the properties of missing elements. After the discovery that some of the elements in the table were out of order, a more reliable table was created by arranging the elements in the order of their increasing atomic number.

A simplified version of the periodic table of the elements, showing only the more common elements, is given in Table 1.2. The adjacent elements in each column have similar properties, which change progressively as you go down the column. Thus the atoms of the elements in group 1A have one valence electron, whereas those in group 2A have two valence electrons. The atoms of the elements in group 7B have seven valence electrons; those in group 8B have eight electrons and are usually inert.

Elements with just a few valence electrons find it easier to lose electrons than to gain enough to complete a stable shell of eight electrons. Elements that readily lose electrons are known as metallic elements. For example, the calcium atom loses its two valence electrons, stripping down to the next stable shell of eight electrons, rather than gaining six electrons. The loss of two negatively charged electrons gives the previous electrically neutral atom a positive charge of $+2$. However, as you move from 3B to 7B, the elements increase their number of valence electrons from 3 to 7 and become increasingly more nonmetallic. For example, chlorine with seven valence electrons can achieve a valence shell of eight electrons more easily by taking an electron from a metallic atom than by gaining a negatively charged electron. This forms a chloride ion with a charge of -1. Neutral atoms which have gained or lost one or more electrons are called ions.

With over 100 different elements it is difficult to learn the properties of each single one. Instead, the periodic table allows you to pick a common element in each group to study as a representative element.

The properties of the elements change gradually in the vertical direction within a group. For example, sulfur is found in group 6B. It forms hydrogen sulfide (H_2S), sulfur dioxide (SO_2), and sulfurous acid (H_2SO_3). The other elements of the group, such as selenium and tellurium, can be predicted to form similar compounds. As another example, consider group 7B, the halogen group. Each one of the halogens forms an acid when it combines with hydrogen. However, at room temperature chlorine is a gas and iodine is a solid. What would you expect bromine to be? It is a liquid.

The transition elements fit in between groups 2A and 1B. These elements had not been discovered at the time Mendeleyev made his historic periodic arrangement of the elements that were known at that time. These important transition elements are characterized by a choice of the number of valence electrons to be used. This occurs as the elements fill in the N and larger electron shells. Their atomic charge and the formula of their ions depend on the reaction conditions.

The first column of elements, group 1A, is called the alkali metals. Their oxides react with water to form strong alkalis. The second column, group 2A, is called the alkaline earth group. Their oxides form moderately alkaline solutions with water. Groups 3A through 8A are the transition elements. These elements have several valences. Group 7B is called the halogen group. Except for fluorine, these halogens make compounds with hydrogen which ionize in water to form strong acids. Group 8B contains the noble gas elements.

From its original eight columns, the periodic table of the elements has appeared in a variety of forms, including a three-dimensional spiral staircase. The modern periodic table has been expanded to 109 confirmed elements. Dr. Glenn T. Seaborg, Nobel Laureate, is the codiscoverer of elements 94 to 102 and 106 (named seaborgium after him) (Fig. 1.3). He is also codiscoverer of nuclear energy isotopes Pu, U, and Np and other isotopes, including I, Fe, Te, and Co. The last few elements in the table are extremely unstable. Barely enough of them have been isolated to verify their position in the table.

1.5 Inorganic Materials

Inorganic chemistry deals with the corrosion of metals and the structures and properties of cement, glass, ceramics, and soils. Consideration of electronic bonding leads directly into inorganic

Figure 1.3 Dr. Glenn T. Seaborg pointing to seaborgium (106) of the periodic table. (*Courtesy of Lawrence Berkeley Laboratory, University of California.*)

chemistry and its many varied applications, such as soils, cement, and metallic corrosion.

1.5.1 Nomenclature

Chemical names are designed to give as much information as possible. For example, the chemical name lithium nitride indicates that lithium, a metal, is combined with nitrogen without oxygen. It says nothing about the stability of the compound. Lithium nitrite has some oxygen in the compound, whereas lithium nitrate has the maximum amount of oxygen with which lithium and nitrogen can normally combine.

If an element has several valence electrons with nearly the same energy, the element has a choice of the number of valence electrons it can use to form a compound. The metals iron and chromium have this multivalent character. Use of a smaller number of valence electrons is shown by the ending "-ous" in the name of the compound, whereas "-ic" is used when a larger number of valence electrons is used. For example, ferrous chloride has $FeCl_2$ for its formula, while ferric chloride has the formula $FeCl_3$. The relative abundance of chlorine determines which of these two compounds is formed. The table of atomic number, mass, and electron shells (Table 1.1) and a partial periodic table of the elements (Table 1.2) are helpful guides in writing formulas of compounds.

1.5.2 Chemistry of aluminum and iron

Silicon is the second most abundant element in the earth's solid crust (28 percent), aluminum is the third most abundant element (8 percent), and iron is next (5 percent). Iron, in use since 1000 B.C., is prepared by heating the iron ore with some form of carbon. Carbon reacts with the iron oxide in the ore to reduce it to iron metal. Corrosion is the natural process that converts iron back into ore.

Although aluminum is more abundant than iron, it was just a century ago that chemists learned to prepare aluminum metal in commercial quantities. Unlike iron, made by heating the ore with carbon, aluminum is made by the electrolysis of a molten mixture of sodium aluminum fluoride and aluminum oxide. The absence of aluminum as a free metal in the earth's crust shows that it is even more chemically reactive than iron. Fortunately, human-made aluminum rapidly forms an invisible oxide coating that tightly covers the metal surface and helps protect it from further oxidation.

Chemists represent the reaction of aluminum with oxygen to form aluminum oxide by the following chemical equation:

$$4Al(s) + 3O_2(g) \rightarrow 2Al_2O_3(s) + \text{heat}$$

This equation shows that four aluminum atoms combine with the six oxygen atoms in the three diatomic oxygen molecules to form two aluminum oxide molecules. The actual molecular formulas are used for elements only if they are gases. The atomic arrangement of atoms in solids and liquids is so much more complicated that it is customary to use the simplest formula possible; for example, Al_2O_3. Letters in parentheses, such as *s, l, g,* and *aq,* are sometimes used to indicate solid, liquid, gas, or dissolved in water. The reactants on the left side of the equation may be separated from the products by either an arrow or an equals sign.

To obey the law that mass can be neither created nor destroyed by ordinary means, an equation must be balanced so that it has the same number of atoms of each kind of element on both sides of the chemical equation. The heat evolved comes from excess chemical energy of the reactants. Reactions that evolve heat are usually spontaneous.

1.5.3 Atomic weight (atomic mass)

Atoms are too small to be weighed individually, but their masses can be compared by using a mass spectrometer. (Note that mass is used instead of weight.) For example, astronauts have mass and substance when in orbit, although they are weightless. Originally the atomic mass of an element was the number of times heavier it was than the hydrogen atom, until hydrogen was found to have three isotopes. It raised the question of which hydrogen isotope to use as the standard. Finally it was decided to use the most common carbon isotope as a standard of exactly 12 atomic mass units. This makes the atomic mass of hydrogen's isotopes average 1.008 atomic mass units. The values given for the elements in Table 1.1 are in atomic mass units.

1.5.4 The mole concept

Biologists can experiment with one small rat or one cell of an organism; chemists usually must be content with a large enough number of atoms or molecules to be able to see them and weigh them. Therefore chemists have adopted Avogadro's number of atoms and of molecules, or whatever structural unit is being considered, as 1 mole (mol) of the substance. Metals, like iron, form metallic rather than molecular structures. 1 mole of iron would weigh 55.85 grams, since the atomic weight of iron in Table 1.1 is 55.85. 1 mole of hydrogen (H_2) weighs 2.0158 grams. This is the molecular weight of hydrogen, which is twice its atomic weight, since there are two atoms of hydrogen in its molecule.

One mole of each of these elements or compounds would contain the same number of chemical units and would be equal to Avogadro's number. This universal constant has been established as 6.02×10^{23}, or 602 followed by 21 zeros.

Similarly, 1 mole of a compound would weigh the number of grams equal to the sum of the atomic masses of the atoms in the formula. Thus for hydrogen chloride (HCl) 1 mole weighs 36.46 grams.

1.5.5 Chemical equations and interpretation

A balanced chemical equation summarizes much information. It predicts the reaction products if the reaction is spontaneous and the amounts produced if the reaction goes to completion. Consider the equation

$$Fe(s) + 2HCl(aq) \rightarrow FeCl_2(aq) + H_2(g) + heat$$

This balanced chemical equation shows that one atom of solid iron reacts with two molecules of hydrochloric acid in water solution to form one molecule of iron chloride in water solution and one molecule of hydrogen gas with the evolution of heat.

The balanced equation also shows that 1 mole, or 55.85 grams, of iron reacts with 2 moles, or 72.92 grams, of hydrochloric acid to form two products: 1 mole, or 126.75 grams, of iron chloride and 1 mole, or 2.016 grams, of hydrogen gas. The sum of the reactants is 128.77 grams. This is the same as that of the products, showing that there is no measurable change in mass during the reaction.

Balanced equations can be used to calculate volumes of gases involved. Experiments have shown that 1 mole of any gas occupies a volume of 22.4 liters (L) measured at standard temperature and pressure (STP), 0°C and 1 atmosphere pressure. The reaction shows that 55.85 grams of iron liberates 1 mole, or 22.4 liters, of hydrogen gas at STP.

Since many reactions between inorganic compounds occur in water solutions, they are usually written as ionic equations. The reaction between iron and hydrochloric acid can be written as

$$Fe + 2H^+ + 2Cl^- = Fe^{++} + 2Cl^- + H_2$$

This form of the equation emphasizes the ionic nature of the reaction. Now there are two chloride ions on each side of the equation that should be canceled to give the skeleton, or net ionic equation, shown:

$$Fe + 2H^+ = Fe^{++} + H_2$$

This ionic equation has the advantage of showing the bare essentials of the reaction.

1.5.6 Chemical equilibrium

In a closed system in which the reactants and the products are confined, experiments have shown that a chemical reaction does not go to completion and stop. In reality, the reverse reaction is also occurring. At equilibrium, the reactants are formed as rapidly as they are used in the reverse reaction. Thus the reaction only appears to stop. To emphasize that an equilibrium has occurred with measurable amounts of each ingredient remaining, arrows going in both directions, or an equals sign, may be used. Chemical equilibrium is an important concept in chemistry.

Reactions that have gone to equilibrium will shift to the left to form more reactants if the amount of one of the products is increased, or go to the right if one of them is decreased. This is known as the mass action rule. How can we apply this rule to the ionization reaction of water to form hydrogen ions? Fortunately, this is a well-understood equilibrium; it is discussed in the next subsection.

1.5.7 Ionic equilibrium in water

The equilibrium reaction of water as it ionizes to form hydrogen ion and hydroxide ion is given by the equation

$$H_2O = H^+ + OH^- - \text{heat}$$

At 25°C (77°F) the product of the hydrogen ion concentration (moles of H^+ in 1 liter) multiplied by the hydroxide ion concentration (moles of OH^- per liter) has been found to equal 1.0×10^{-14}. In pure water the two ions are produced in equal amounts or concentrations. The square root of the ionization constant is 1.0×10^{-7} (one ten millionth). This means the hydrogen ion concentration in the 1 M (1 mole per liter) solution used for the hydrogen electrode is 10 million times larger than the hydrogen ion concentration in pure water. This is a common experience for the chemist who, when referring to hydrogen ion concentration, defines pH as the negative exponent of 10. In other words, the exponent of 10 without the minus sign is known as the pH. This is illustrated in Fig. 1.4.

If the pH is not a whole number, it can be calculated from the equation

$$pH = -\log(H^+)$$

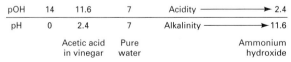

Figure 1.4 pH scale for typical water solutions.

In earlier examples the log of 1.0 is zero, and the log of 10 to a negative exponent is equal to the exponent itself. Thus a whole number was obtained for each pH of a dilute acid. Notice that at room temperature the sum of the pH and the pOH in water will always be 14.00 by this convention.

From Fig. 1.4 it can be seen that whether a substance is acid or alkaline is a matter of degree. A pH smaller than 7 is acid; a pH larger than 7 is alkaline. Water from rain is usually slightly acid due to products of combustion and other pollutants in the air. One of these is carbon dioxide, which is so constant in air that it helps to establish the pH of water in swimming pools.

Acids which ionize almost completely are called strong acids and include hydrochloric, nitric, sulfuric, and the first hydrogen of phosphoric acid. Strong bases also ionize almost completely and include the hydroxides of the metals of groups 1 and 2 in the periodic table. Weak acids that ionize only 5 or 10 percent are common and include acetic acid (HOAc), carbonic acid, and the second and third hydrogen to ionize from phosphoric acid.

The most common weak base is ammonium hydroxide. Although all soluble salts ionize almost completely, the ions of the salts of a weak base or a weak acid react with water, or hydrolyze, as shown by the equation for the reaction of the acetate ion with water,

$$OAc^- + H_2O = HOAc + OH^-$$

where OAc^- is an abbreviation for the acetate ion.

Note from this reaction that the acetate ion reacts with water (hydrolyzes) to form an alkaline solution. Actually it has a pOH of 4.6 and a pH of 9.4, which is far from neutral. Also, a solution of 1 M ammonium chloride has a pH of 4.6. For comparison, lemon juice has a pH of about 2.3, vinegar about 2.5, and orange juice about 3.5. A 1 M solution of sodium carbonate has a pOH of 12.1. With a pH of 0.6, a solution of trisodium phosphate (TSP) was used as a heavy-duty alkaline cleaning compound until nondegradable phosphate increased in streams and lakes and became a serious pollution problem.

1.6 Organic Materials

Metals, wood products, and other common organic building materials are rapidly being replaced by synthetic organic materials. Extensive research, development, and promotion of synthetic materials have made a basic knowledge of organic chemistry an invaluable aid to understanding the best uses of synthetic materials.

1.6.1 The synthetic world of organic chemistry

The clothes we wear, cars we drive, much of the food we eat, the furnishings in our homes, and the materials of construction and their protective coatings are new to our generation. The alchemist's dream to make gold, diamonds, sapphires, and much more has come true. But what may be a delight to some of us is a source of confusion to others. In this exciting age, our choices are becoming more difficult. How is one to take advantage of such rapid progress? We should ask searching questions and learn the basic concepts and the language of organic chemistry.

1.6.2 Scientific language

The arts and humanities have their own languages, so why not science? An understanding of the language of chemistry is useful and needed so we can distinguish from one another the many chemical compounds making up our building materials. For example, chemistry describes the preparation, properties, and structure of organic and inorganic materials. But what do we mean by organic in contrast to inorganic materials?

The dictionary may describe an organic material as any substance of animal or vegetable origin. Inorganic, or not organic, refers to chemicals from mineral sources. The chemical symbol for an element is an abbreviation of its name. The first letter of the Latin, Greek, or English name of an element is often used as a symbol when writing chemical formulas. To avoid confusion, the second or third letter of the name may be used. Chemical combinations of elements are called compounds.

It is customary, when writing formulas for inorganic compounds, to put the symbol for the metal first, followed by the symbol for the nonmetal. The common termination "-ide" in the name shows that no oxygen atoms are contained in the compound. The presence of one atom of oxygen is shown in the molecular formula by changing the termination to "-ite." The termination "-ate" means that more than one oxygen atom is contained in the compound.

The prefix "per," as found in sodium perchlorate, is reserved for an unusually large number of oxygen atoms in the molecule. Compounds with per in their name are usually unstable and explode if they come in contact with substances that react with oxygen. For example, about a gallon of perchloric acid ($HClO_4$) was often used in an electroplating bath in contact with a plastic basket. One day such a bath exploded without warning. This explosion, occurring in a Los Angeles laboratory in the 1950s, was so violent that an entire city block of buildings was demolished. What happened? The excess oxygen in the perchloric acid reacted violently with the carbon and hydrogen in the plastic to form carbon monoxide and water vapor. The technician was vaporized and never found.

Sodium metal readily reacts with the nonmetal chlorine to form a salt called sodium chloride. The chemical compounds sodium chloride, sodium chlorite, sodium chlorate, and sodium perchlorate are represented by the formulas NaCl, NaClO, $NaClO_3$, and $NaClO_4$, respectively. Na is the abbreviation for natrium, the Latin name for sodium. Cl stands for chlorine, and O stands for oxygen. The subscripts 3 and 4 indicate three and four atoms of oxygen, respectively, in the molecule. Other metals and nonmetals also react to form salts. The term "salt" is not restricted to sodium chloride, ordinary table salt.

When oxygen reacts with metals, such as sodium or calcium, the compounds formed are called oxides. Oxides react with water to form hydroxides, also known as bases or alkalies. Calcium oxide has the formula CaO, is known as lime, and reacts with water to form $Ca(OH)_2$, a base. Bases have the common property of feeling slippery to the touch and tasting bitter. Bases can be recognized by their name ending with hydroxide, such as sodium hydroxide. Bases react with acids to form salts. The H from an acid combines with the OH of a base to form water (H_2O). An acid formula starts with H, the symbol for hydrogen. An acid can also be recognized by the "-ic" termination in the name followed by the word acid. As an example, hydrochloric acid is known in industry as muriatic acid and has the chemical formula HCl.

1.6.3 The language of organic chemistry

Organic chemistry is concerned with the compounds containing carbon, in contrast to inorganic (nonorganic) chemistry. Carbon, with its four valence electrons, can accept four more electrons from other atoms to form an octet of four electron-pair bonds. The existence of the variety of organic compounds is possible because the carbon atoms can form stable electron-pair bonds with one another either in beadlike chains of atoms or in ring structures. The study of these beadlike chains of carbon atoms is called aliphatic chemistry. The study of ring structures is called aromatic organic chemistry because of their characteristic smell.

There are probably 20 times as many carbon compounds as there are compounds formed by all the other atoms in the periodic table. This is illustrated by the variety of organic compounds formed by the addition of more carbon atoms to the chain, as shown in Fig. 1.5.

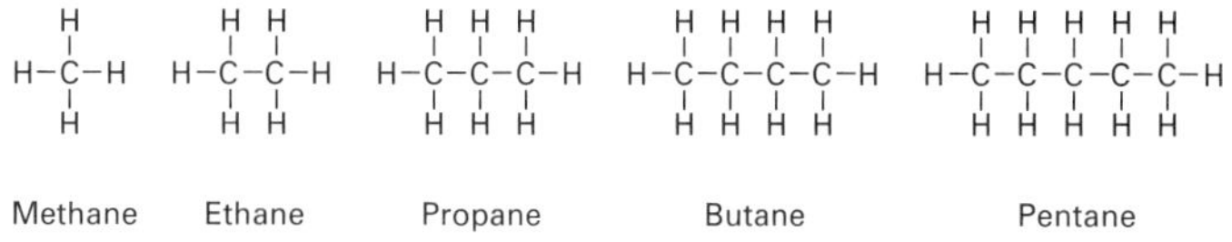

Introduction to Science and Technology 21

```
    H           H H         H H H        H H H H       H H H H H
    |           | |         | | |        | | | |       | | | | |
  H-C-H       H-C-C-H     H-C-C-C-H    H-C-C-C-C-H   H-C-C-C-C-C-H
    |           | |         | | |        | | | |       | | | | |
    H           H H         H H H        H H H H       H H H H H

  Methane      Ethane       Propane      Butane         Pentane
```

Figure 1.5 Aliphatic saturated hydrocarbon series.

The straight-chain aliphatic hydrocarbons, saturated with hydrogen atoms, have the general formula C_nH_{2n+2}, where n stands for the number of carbon atoms. The "-ane" termination in their names indicates a member of the saturated aliphatic series. The first four were given names before the need for a systematic nomenclature was realized. Latin names are used for the number of carbon atoms in the series, starting with five carbon atoms, that is, pent-, hept-, hex-, oct-, non-, and dec-. These form what is called a homologous series because of their gradual change in physical properties, such as freezing and boiling point, due to a structural difference of only CH_2 groups. Beginning with four carbon atoms, forked chains can occur and form two or more different compounds with the same number of carbon atoms, called isomers.

If an electron-pair bond joining two carbon atoms together is replaced by two pairs of electrons, it is called a double bond. The end of the common name is changed from "-ane" to "-ene," as in ethylene, propylene, and butylene. Each carbon atom with a double bond counts the four electrons of the bond as part of its octet. Also each of the two carbon atoms joined by a double bond has one less bond to hold a hydrogen atom. With the loss of two hydrogen atoms, the compound is an unsaturated hydrocarbon. These double bonds are unstable and will take up oxygen from the air to become rancid. Such double bonds are found in the liquid unsaturated fats in our food. (By using a catalyst, the oils can be hydrogenated. This process thickens the oils to form partially hydrogenated, saturated fats, such as margarines, which are solid at room temperature.)

These unsaturated compounds are known as alkenes or olefins. Their general formula becomes C_nH_{2n}. Since a single chemical bond is indicated by either two dots or a single dash, a double bond needs four electron dots or two dashes, and a triple bond is indicated by three dashes to form the alkynes, as shown in Fig. 1.6.

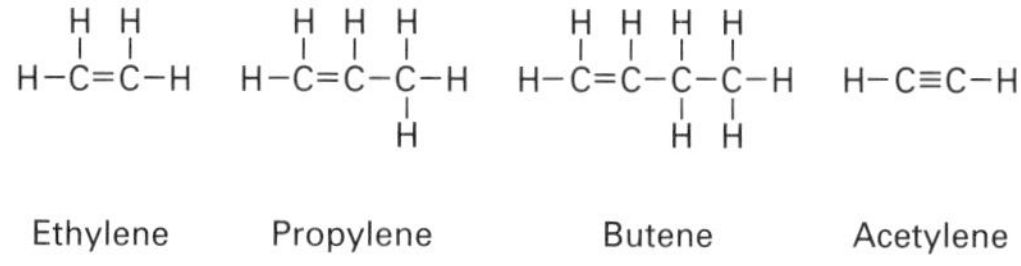

```
   H H         H H H        H H H H
   | |         | | |        | | | |
  H-C=C-H    H-C=C-C-H    H-C=C-C-C-H     H-C≡C-H
               |            | |
               H            H H

  Ethylene    Propylene      Butene        Acetylene
```

Figure 1.6 Unsaturated alkyne hydrocarbons.

1.6.4 Aliphatic hydrocarbons with other elements

If one of the hydrogen atoms is replaced with an –OH (or hydroxyl group), an inorganic alkali does not result. Instead, a class of compounds called alcohols is formed, as illustrated in Fig. 1.7.

Methyl alcohol is also known as methanol, or wood alcohol, since it was made by the destructive distillation of wood before it was produced synthetically. Ethyl alcohol, made by the fermentation of grain, is known as ethanol, or grain alcohol. Wood alcohol causes blindness if a human drinks it.

The hydrogen atom of a hydrocarbon can be replaced by a halogen such as fluorine, chlorine, bromine, or iodine. For example, chlorine reacts with methane to form methyl chloride and hydrogen chloride. Ethane can form ethyl chloride. Chloroform is formed if only three hydrogen atoms are replaced by chlorine. If all four hydrogens on methane are replaced by chlorine atoms, carbon tetrachloride is formed. Carbon tetrachloride, a toxic chemical, is banned from public use.

Other chemical modifications of aliphatic hydrocarbons are illustrated in Fig. 1.8. When the R– symbols in the general structural formulas at the top of the figure are replaced by the radicals listed on the right below, the organic compounds are formed as named.

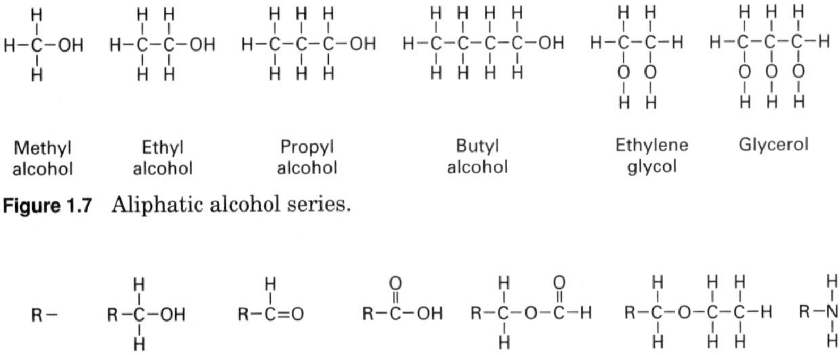

Methyl alcohol | Ethyl alcohol | Propyl alcohol | Butyl alcohol | Ethylene glycol | Glycerol

Figure 1.7 Aliphatic alcohol series.

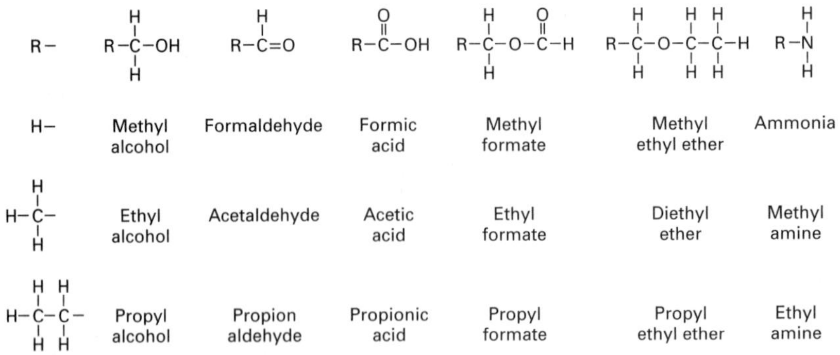

Figure 1.8 Oxygen-containing aliphatic compounds.

It is convenient to divide an organic compound into two parts. One part is the reactive, functional group, and the other part is the radical group symbolized by the letter R. For example, an alcohol becomes ROH. Replacing the R– by a methyl radical gives methyl alcohol. Replacing R– by an ethyl radical gives ethyl alcohol.

The acids, bases, and salts of inorganic chemistry have their counterparts in organic chemistry. An organic radical attached to the –COOH carboxyl group corresponds to the inorganic acid. An organic amine, such as methyl amine, a derivative of ammonia, is the equivalent of an inorganic base. This typical amine reacts with hydrochloric acid to form the salt, methyl amine hydrochloride, similar to ammonium chloride.

The organic acid functional group is –COOH. Thus if R– is a hydrogen radical, its combination with the acid functional group, or radical, gives HCOOH, or formic acid. If the R– stands for the methyl radical, the acid becomes acetic acid. The organic amine has the –NH_2 functional group. If R– is H–, it becomes ammonia.

Many of the compounds listed in Fig. 1.8 are hazardous, if not toxic. In fact, it is a general rule that any chemical taken in excess can be poisonous to the body. For example, formic acid is the poison in bee stings. Diethyl ether, made by dehydrating ethyl alcohol, is an anesthetic, but is no longer used because of the danger of an explosion. Anything in excess, even salt, can be poisonous.

1.6.5 Aromatic series of carbon compounds

Ring structures and, in particular, those containing benzene rings typify the aromatic compounds. Benzene has only enough bonding electrons for three double and three single bonds. The reactivity of the bonds indicates that all six carbon atoms in the ring share the electrons equally. In drawing aromatic structures, it is customary to either alternate double bonds with single bonds between carbon atoms in the ring, to use a circle in the center of a hexagon, or simply to use a hexagon. This benzene structure is shown in Fig. 1.9 in the formulas for xylene, aniline, and styrene.

The simplest aromatic compound is benzene itself (C_6H_6), which is flammable and toxic. Many other aromatic compounds contain the benzene ring structure. As shown in Fig. 1.9, the benzene radical, called phenyl, is drawn as a hexagon without showing the carbon or hydrogen atoms.

The functional group –OH attached to the phenyl radical gives it the property of an acid. Phenol is a weak acid which ionizes so little that it provides only a small amount of hydrogen ions in water solu-

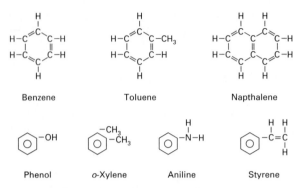

Figure 1.9 Common aromatic compounds.

tion. Most organic compounds are only slightly soluble in water, if at all, but the solubility is increased by the presence of a hydroxyl, amine, or similar functional group. Reactions between organic chemicals, in general, are much slower than those between inorganic ions in solution. Where several different organic products are possible, the product desired can be favored by the selection of the catalyst or other reaction conditions.

Aniline has the amine functional group and corresponds to the inorganic "alkali." Like other amines, it reacts with hydrochloric acid to form a saltlike water-soluble compound, called aniline hydrochloride.

1.7 Common Synthetic Polymer Types

Not only do carbon atoms join together in nature to form moderate-sized chain and ring compounds, but some natural compounds, such as rubber, consist of enormously long molecules called polymers. Other natural polymers contain oxygen and carbon atoms in their chain (cellulose) or nitrogen and carbon (protein). Chemists also have learned to synthesize many varieties of organic polymers which are used in paints, plastics, adhesives, and construction products, such as polymer-impregnated concrete.

At one time the automobile industry was dependent on natural rubber from the saplike latex of rubber trees in southeast Asia. Cut off from their supply by World War II, chemists were challenged to synthesize a rubber substitute. Not only were chemists successful in making rubber, but they improved upon nature with varieties which are more oxidation-resistant and wear longer.

A monomer is a small, reactive organic molecule that can combine with other monomers to form a macromolecule called a polymer. In other words, organic polymers are large molecules (macromolecules) formed from monomers. These monomers are small reactive organic molecules which can combine with one another. Often polymers are classified by how they behave when heated. If they soften when warmed, they are called thermoplastic polymers. However, if they never become soft but only char after intensive heating, they are known as thermosetting polymers or thermosets.

1.7.1 Thermoplastic polymers

Thermoplastic polymers, such as vinyls, are made by joining monomers together to form long chains. This can be done by means of a catalyst, such as benzoyl peroxide. The catalyst causes one of the bonds in the double bond in the monomer to open and form a free radical. A free radical is an unstable, reactive kind of a molecule. In this polymerization reaction these radicals link together to form long chains of repeated polymeric units. Each molecular chain may have a molecular weight of from several thousands to a million.

In Figs. 1.10, 1.11, and 1.12 the dashes on either end of the polymeric units represent electron-pair bonds. These bonds connect adjacent polymeric units (not shown) together. The brackets deliberately cut the dashes into two pieces. This is to show that half of the bond (one electron) came from the adjacent polymeric unit (not shown). The degree of polymerization is represented by the letter n, which stands for the number of polymeric units in the polymer.

Figure 1.10 shows the polymerization of vinyl chloride to form the poly(vinyl chloride) polymer. As illustrated in this figure, when one of the electron-pair bonds (of the double bond of the vinyl monomer) is broken, one electron goes to each end of the monomer, forming a free radical. Each of these electrons might be pictured as an electron "hand" groping for the free electron hand of another monomer radical. The clasped hands represent a new electron-pair bond that holds the two radicals together. Thus monomer radicals become joined together to form a chain.

Usually the polymer chains grow in an orderly fashion. Some of the chains have branches or become forked and grow into a tangled mass rather than a very long chain.

To limit the chains to the desired length, monomers with only one functional group (chain stoppers) are used in small amounts to terminate any chain they join.

$$\begin{array}{c}H\\|\\C\\|\\H\end{array} = \begin{array}{c}Cl\\|\\C\\|\\H\end{array} + \begin{array}{c}H\\|\\C\\|\\H\end{array} = \begin{array}{c}Cl\\|\\C\\|\\H\end{array} + \text{catalyst} \longrightarrow \left[\begin{array}{cccc}H & H & H & Cl\\|&|&|&|\\C & - & C & - & C & - & C\\|&|&|&|\\H & Cl & H & H\end{array}\right]_n$$

Figure 1.10 Polymerization of vinyl chloride.

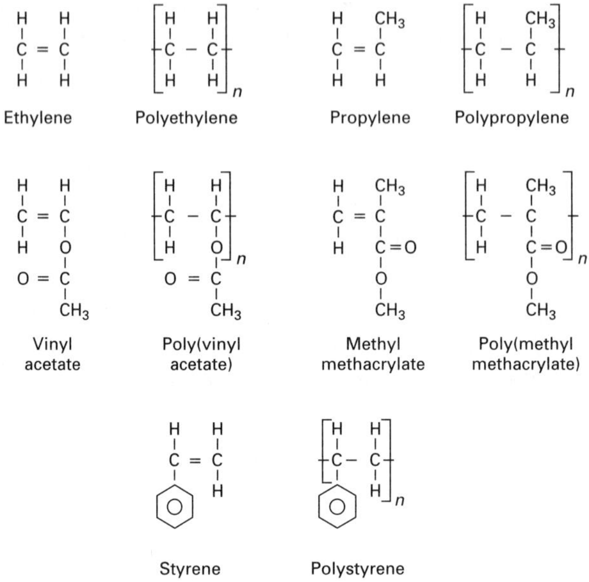

Ethylene / Polyethylene / Propylene / Polypropylene

$$\begin{array}{c} H \quad H \\ | \quad | \\ C = C \\ | \quad | \\ H \quad H \end{array} \qquad \left[\begin{array}{c} H \quad H \\ | \quad | \\ C - C \\ | \quad | \\ H \quad H \end{array}\right]_n \qquad \begin{array}{c} H \quad CH_3 \\ | \quad | \\ C = C \\ | \quad | \\ H \quad H \end{array} \qquad \left[\begin{array}{c} H \quad CH_3 \\ | \quad | \\ C - C \\ | \quad | \\ H \quad H \end{array}\right]_n$$

Ethylene   Polyethylene   Propylene   Polypropylene

Vinyl acetate   Poly(vinyl acetate)   Methyl methacrylate   Poly(methyl methacrylate)

Styrene   Polystyrene

Figure 1.11 Common polymers and their monomers.

Many thermoplastic polymers can be decomposed if they are heated to a high temperature because a monomer unit breaks off. Most of these monomers have a distinctive aroma. This is one way to identify a polymer.

Structural diagrams of other monomers which join to form thermoplastic polymers are shown in Fig. 1.11. These unite with one another, stepwise, to form the chain of carbon atoms known as addition polymers or chain-growth polymers. The n shown in Fig. 1.11 indicates the number of repeating units in the final chain.

Each of these polymers can be made to almost any specification that is needed. Polyethylene is an illustration. As normally polymerized, ethylene forms a polymer with tangled much-branched chains which are not well-packed. This form of the polymer is known as low-density polyethylene. With different catalysts, ethylene can be polymerized to form a polymer with straighter and more closely aligned chains, which, being more closely packed together, are more dense. This high-density form of polyethylene has a higher melting point and more than double the tensile strength of the low-density polymer.

Whether a monomer would be predicted to form a thermoplastic or a thermosetting polymer depends on the number of double bonds or functional groups in the monomer. One double bond or two functional groups in a monomer can produce a thermoplastic (linear) polymer. If long enough, these linear chains form a solid polymer which, when heated, will soften and flow as the chains slide over one another.

1.7.2 Thermosetting or cross-linked polymers

Thermosetting polymers, formed by monomers with two or more double bonds or functional groups, form cross-linkage between chains. These polymers, unlike the previous thermosetting polymers, do not soften, but they do char after intensive heating.

Figure 1.12 shows the formation of a phenolic thermosetting resin. The reaction also represents a condensation polymerization reaction since the two chemicals react together (condense) to form the start of a polymer and split off a small molecule, in this case water.

Made from phenol and formaldehyde, this phenolic (phenol formaldehyde) resin was discovered by Bakeland in 1907 and named Bakelite. It is a good illustration of a common three-dimensional cross-linked thermosetting polymer.

Another important thermosetting polymer (resin) is the epoxy class of polymers. The epoxy (epoxide) resin is well established in the building industry. This resin is used as an adhesive, a protective coating, and a general-purpose bonding agent.

Progress in the development of new types of polymers has been rapid in the last few decades. For example, a mixture of two different monomers polymerizes to form a polymer, called a copolymer. Glass fibers can be added to monomers to form a reinforced polymer (composite) with increased strength. Research in the nature and methods of formation of old and new composites has expanded to concrete, mortar, metals, glass ceramics, and polymers. As described in later chapters, composites offer the promise of new and improved properties for common building materials.

Chemistry is primarily an experimental science. Chemists have developed many theories. Some of the theories have been experimentally verified beyond a reasonable doubt and have become laws. Even these may be modified by more precise experiments performed with improved instruments or when applied to unusual situations. A unique feature of an experimental science is that each scientist veri-

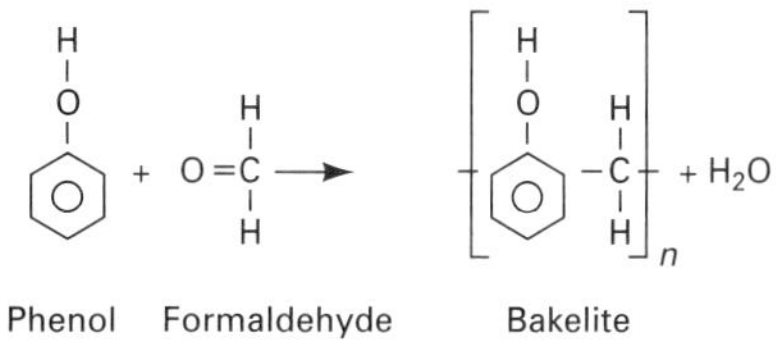

Figure 1.12 Phenolic resin formation.

fies, or adds to, our scientific knowledge. Very little of the information is discarded, only modified and expanded.

Terms Introduced

Acidity pH
Acids
Addition polymer
Alcohols
Aliphatic hydrocarbons
Alkalies
Aluminum oxide
Atom
Atomic mass
Atomic number
Atomic weight
Avogadro's number
Bases
Carbon tetrachloride
Catalyst
Chemical equilibrium
Chemical symbol
Compounds
Condensation polymerization
Covalent
Double bond
Electron
Electron bond
Electron shells
Electron spin
Element
Epoxy
Ether

First law of thermodynamics
Formic acid
Free radical
Functional group
Hydrolyze
Ionic bond
Ionize
Ions
Isomers
Isotope
Mass action rule
Mendeleyev
Mole
Monomer
Net ionic equation
Neutrons
Nucleus
Octet electronic structure
Organic chemistry
Oxides
Periodic table
Phenyl
Polymer
Polymeric units
Protons
Seaborgium
Second law of thermodynamics
Shared electron bond

Strong acids
Thermoplastic polymer
Thermosetting polymers
Transition elements

Triple bond
Unsaturated hydrocarbon
Valence electrons
Weak acids

Questions and Problems

1.1. Models are useful. They help you to visualize theories and phenomena and to predict the results of their applications to new situations. Explain how water can evaporate into the air from a latex paint without the water being heated to the boiling point.

1.2. Which one of the lead compounds contains no oxygen: lead chlorate, lead chloride, lead chlorite, or lead perchlorate? Which one contains the most oxygen?

1.3. Predict whether the radioactive gas radon in column 8B of the periodic table of the elements, Table 1.2, would be expected to form compounds with other elements. Radon is reported to occur in the basements of some homes in concentrations high enough to produce cancer.

1.4. Interpret from their pH which of the following solutions are acidic and which are basic: (*a*) 7.50 pH faucet water; (*b*) 6.75 pH distilled water; (*c*) orange juice; (*d*) lemon juice.

1.5. Distinguish between organic and inorganic materials and give an example of each.

1.6. Distinguish between aliphatic and aromatic compounds and give an example of each.

Suggestions for Further Reading

Gessner G. Hawley, *Condensed Chemical Dictionary,* 12th ed., Van Nostrand Reinhold, New York, 1993.
Wesley E. Lingren, *Essentials of Chemistry,* Prentice-Hall, Englewood Cliffs, N.J., 1986.
McGraw-Hill Encyclopedia of Chemistry, 2d ed., McGraw-Hill, New York, 1993.
Sybil P. Parker, ed., *McGraw-Hill Dictionary of Scientific and Technical Terms,* 5th ed., McGraw-Hill, New York, 1994.

Chapter

2

Site Investigation, Preparation, and Environment

2.1 Introduction	31
2.2 Classification of Soils	32
2.3 Site Investigation	33
2.4 Soil Structure, Composition, and Behavior	34
2.5 Structure of Clay	35
2.6 Site Preparation for Structural Foundations	37
2.6.1 Soil Drainage	39
2.6.2 Compaction Methods	40
2.6.3 Chemical Treatment of Clay Soils	40
2.6.4 Piles and Caissons	41
2.7 Our Environmental Responsibilities	44
2.7.1 Recycling Used Building Materials	44
2.7.2 Occupational Safety and Health Administration	45
2.7.3 Environmental Impact Statement	46
2.7.4 Environmental Response, Compensation, and Liability	46
2.7.5 Structural Environmental Hazards	48
2.8 Soil as a Growth Medium	48
2.8.1 The Physical Nature of Soils	49
2.8.2 Permeability of Soil to Water and Nutrients	50
2.8.3 Major Soil Nutrients	51
2.8.4 Trace Soil Nutrients	53
2.9 Hazards of Pesticide Usage	55
2.10 Irrigation and Industrial Waste Pollution	56
2.11 Environmental Considerations	57
Terms Introduced	58
Questions and Problems	58
Suggestions for Further Reading	59

2.1 Introduction

This chapter considers soil classification and properties, site investigation, foundations of buildings, environmental responsibilities, recycling of building materials, and landscaping. Building site preparation includes drainage, soil compaction, soil treatment, and use of piles and caissons. Because of the ever-increasing relevance to the environment, there is a discussion of the recycling of usable building materials and the concerns of the Environmental Protection Agency (EPA), environmental impact statements (EISs), the Comprehensive Environmental Response, Compensation, and Liability Act (CERLA), and the Occupational Safety and Health Administration (OSHA). This is followed by the effects of landscaping on the building site and on the environment.

2.2 Classification of Soils

From their origin as igneous and metamorphic rock, all soils are produced in nature by weathering, glacial action, earthquakes, and sedimentation. Weathering can be mechanical or chemical. Mechanical action involves large rocks which are crumbled by massive movements of glaciers over the earth's surface and earthquakes in the earth's crust, and by rain, wind, and freezing temperatures. The chemical action of oxygen, water vapor, and carbon dioxide in the atmosphere eventually erodes the rocks into fine particles, chemically modifying the soil.

Soil formed by glacial action is known as till and soil deposited by rivers and streams is called aluvian. Sedimentary soil and rock is that which has settled out from rivers, lakes, or ancient seas.

Soil texture is classified as gravel, sand, silt, or clay. This classification can be determined by measuring the diameter of the particles: very coarse sand (about 2 millimeters), fine sand (0.2 millimeters), silt (0.02 millimeters), and clay (0.002 millimeters and smaller). Clay particles can remain suspended in rivers for extended periods of time. In more concentrated suspensions, clay can form plastic clay soil deposits.

Soils can be grouped by their chemical composition, such as insoluble compounds of oxides, carbonates, and sulfates, and silicates of calcium, magnesium, aluminum, and iron. Some common examples are marble and limestone (calcium carbonate) and gypsum (calcium sulfate). Other examples include mineral mixtures, such as granite (crystals of feldspar and quartz) and feldspar (potassium aluminum silicate). Clays are more complex aluminum silicates containing sodium, calcium, magnesium, or other cations to form a variety of clay minerals. Of particular interest is the instability of building foundations supported by the ground underneath containing colloidal clay minerals.

The ground structure as well as the types of soil beneath a building site are important to the design of a foundation. Indications of fissures in the soil structure, underground water passages, the water table, and water seepage are all matters of concern.

2.3 Site Investigation

The purpose of a thorough site investigation is to assess the suitability of the area for the proposed construction, to suggest ways to improve the load-bearing ability of the subsoil, to guide the design of an economical and safe foundation, to foresee construction problems related to ground conditions, and to call attention to any adverse effects on adjoining structures. A survey by a soils engineer is essential. This represents less than 1 percent of the total cost of the building, but is probably one of the most critical factors in its long-term performance.

Preliminary to the site and soil investigation, the engineer should submit a proposal including the number of boreholes planned and their locations and depths. Sampling procedures for laboratory tests should be included. The owner or developer of the property and the engineer should decide whether an interpretative report and a foundation design are to be included in the final report.

Soil profiles should show the sequence of strata, their locations, their nature, and the thickness and structure of the undisturbed rock and soil formations. Some soil disturbance and compression is unavoidable, but the soil samples collected for laboratory testing should represent the undisturbed condition as closely as possible. These vertical profiles are needed from several locations on the building site to determine the degree of uniformity and tilt of the load-bearing stratum. For a multistory building the soil should be investigated to a depth equal to the width of the foundation, especially if underground water or extensive clay formations are suspected.

Representative core samples are needed for inspection at the surface and for laboratory testing. Many types of drilling equipment are available to help preserve the quality of core sections for sophisticated laboratory tests. Special skills are needed to remove the samples from the boring tubes in order to obtain undisturbed samples. Nondestructive x-ray radiography is often useful at the building site to search the cores for indications of fissures, fractures, and other irregularities in soil structures.

The final report should contain important information, including a geological description of the region. A sketch of the building site should provide the locations and depths of the boreholes. The soil profile will show the location and thickness of each stratum encountered, the kinds of soil contained, and the soil structure. A full report of the tests made on the laboratory samples will include the results and their interpretation. This report should also give data on soil compressibility, shear strength, size distribution of sand particles, clay plasticity of the stratum samples, location of any subsoil cracks or fissures, moist areas, and water tables. Any construction problems related to soil conditions should be anticipated and suggestions given for their prevention.

The conclusions should describe any required soil drainage, soil strength improvement, or chemical treatment. Soil improvement could mean soil compaction by temporarily loading the building site with rocks, soil, or heavy equipment. Treatment could mean mixing chemicals with the soil to improve its load-bearing properties. Or piles may be recommended to reach deeper load-bearing strata. Finally, for the most economical and safest foundation, the report should recommend the appropriate type and design of the foundation and method of construction. If indicated, alternative designs of the foundation should be given in detail.

Soils have a complex composition and structure. In order to appreciate and to make proper use of the site investigation report, it is important that the individuals involved have an understanding of soil structure, composition, and behavior under stress.

2.4 Soil Structure, Composition, and Behavior

The design of a foundation requires a thorough knowledge of the methods of soil stabilization as well as an understanding of soil mechanics. Under the heavy load of a structure and its occupants, unstable soil can settle unevenly and cause the building to tilt or slide. Loose soils beneath a foundation can shift and settle. Excavations adjacent to the building can affect water drainage and cause the foundation to tilt.

The field of soil mechanics is the prediction of soil behavior under internal and external forces. One of its branches is the rheology of soils: the study and treatment of the rate of movement of soil under stress or shear. Sandy soils are not only porous to air and moisture, they are also mobile. In some areas the movement of sandy beaches during violent storms and persistent tidal movements is a continual threat. Soils can move from the pressure of a foundation. A change in the drainage pattern of water can remove soil by erosion. And under certain pH and moisture conditions, some clay soils act as a thixotropic gel-like solid.

Building foundations can be endangered by several soil stratum conditions. These include excessive porosity to water, low shear strength, lack of compactness, and clay content. The porosity and low compaction of topsoil permits water to filter down to reach the stratum that is high in clay content. Water lubricates the clay masses, permitting shear (slides). A change in water content can cause soil to swell or shrink. Changes in the subsoil water content can be due to a reduction in exposure of the building site to rain, changes in drainage, or landscape irrigation.

Unstable soils can have fluidlike properties, depending on soil type, particle size, pH, and moisture content. Fluidity can be due to the tendency of the strata of fine sand, silt, or clay to flow (or slide) into cracks and fissures, if stressed. This is more often due to the colloidal nature of extremely fine clay soil particles.

Thixotropic clay in soils beneath building sites can cause slides, erosion, and uneven settling of foundations. For example, soils containing bentonite, a common component of clay, are thixotropic. Thixotropic substances swell when soaked in water. Their viscosity is lowered by stirring. The particles return to a rigid gel structure when stirring stops. The stress of stirring orients the platelets so they can slide past one another more easily, resulting in the lower viscosity.

Thixotropy, a valuable property of paints, prevents the paint from running or sagging when applied to vertical surfaces. This is a necessary property of ceramic materials, permitting them to be shaped until fired. But soils containing enough clay to be thixotropic can produce disastrous mud slides. These slides are often triggered by the alkali in the ashes of a brush fire followed by heavy rains.

2.5 Structure of Clay

Clay is a fine, hydrated aluminum silicate that occurs in varying amounts in soils. Some clay minerals are less thixotropic than others, although they are all essentially hydrated aluminum silicates. Clays differ in the relative number of layers of aluminum oxide (alumina) and silicon dioxide (silica). They also differ in the amount of adsorbed ions, such as the ions of sodium or calcium. Formed by the weathering of rocks, the particular clay mineral that is formed depends on the reaction with the other chemicals present. Of special interest is the difference in the load-bearing capacities of sodium bentonite and calcium bentonite.

Clay minerals consist of layers (sheets) of silica and layers of alumina stacked on top of one another. The silica sheet is made of oxygen tetrahedra (oxygen ion pyramids). The triangular-shaped oxygen bases of the tetrahedra are tightly bonded together to form a sheet. The fourth oxygen of the tetrahedron sticks out above (or below) the sheet. A silicon ion is in the center of each tetrahedron.

The alumina sheets are made in a similar arrangement of oxygen octahedra with an aluminum ion in the center. These sheets are loosely held together by hydrogen bonds. In this case a hydrogen atom that is attached to an exposed oxygen atom from adjacent sheets is said to form a hydrogen bond. This forms many oxygen-hydrogen-oxygen bonds to hold the sheets together. These layers of oxygen ions are held together so loosely that a water molecule can squeeze in between the layers, causing the structure to swell.

Bentonite particles have two sheets of silica sandwiched between single alumina sheets. The exposed oxygen atoms on the crystal faces and edges give these clay crystals a negative charge. This charge is partially neutralized by the adsorption of a blanket of positively charged ions, such as sodium or calcium ions, to form sodium or calcium aluminum silicate crystals.

The looseness with which this swarm of adsorbed ions is held makes it possible to exchange one type of positive ion for another. This ion-exchange process is applied in commercial water-demineralizing systems, so why not use it to convert the sodium clay in soils to the more load-supporting calcium clay soil?

In western clay the holes in the clay crystals can exchange positive ions with those in abundance in the wet soil by a process called ion exchange. In clay laid down by sedimentation from salty water, sodium ions, in high concentration around the clay particles, can squeeze into a hole if a hydrogen ion vibrates too far out of the hole.

Absorption of water by dry clay can cause the soil to swell as much as 10 percent and to have thixotropic properties unless stabilized by the addition of a sodium or calcium salt. Some clays, especially the sodium bentonite type (Wyoming or western type), form a plastic moldable mass with water, becoming a thixotropic gel-like solid. This gel liquefies on agitation or under a shearing pressure, but returns to a stiff gel when the action stops. By contrast, the calcium bentonite (southern) clay has negligible swelling gel-like properties.

Since both kinds of bentonites have strong ion-exchange properties, sodium hydroxide (soda ash) in the ashes of a brush or forest fire could produce the sodium-type clay. Treating the soil with calcium hydroxide (slaked lime) could convert bentonite into the more stable calcium-type soil.

2.6 Site Preparation for Structural Foundations

Foundation types fall roughly into two classes: shallow foundations for homes (Fig. 2.1) or small buildings (Fig. 2.2) and deep foundations for multistory buildings (Fig. 2.3). Shallow foundations usually go below the soil surface a distance less than the shortest dimension of the foundation. In freeze-thaw regions, foundations must go down into the soil below the freeze area to avoid serious deterioration. Deep foundations use groups of poles to transfer structural loads to bedrock or to load-bearing soil beneath the foundation.

In practice, foundations fail mainly because of erosion, settling, sliding down slopes, tilting, or earthquakes. Also, changes in soil volume beneath a foundation can cause serious structural damage due to swelling of soil as it absorbs moisture or when the wet soil expands on freezing. In addition to the effect of wet and dry weather on the soil

Figure 2.1 Concrete footing for small home. A footing of poured concrete supports one corner of a galvanized-steel frame residence. Self-tapping Phillips-head screws and self-tapping hex-head bolts secure the steel structure to the foundation.

Figure 2.2 Site preparation.

Figure 2.3 Foundation preparation for tall building. Site shows grading and footings, housing for storm drain, and utility covers. The concrete block holding the crane will be the elevator foundation. (*Courtesy of University of Hawaii, UH Relations.*)

moisture content, there is a change in the moisture content when grass and shrubbery are planted near a building's foundation.

The three most common methods of improving the compactness and structure of topsoil are drainage, compaction, and chemical treatment. Subsoil at the building site can be modified to increase the total load-bearing ability of the foundations by a variety of methods, depending on the depth of the faulty soil. If the underlying strata are weak and unreliable, then the more costly piled foundation is necessary.

2.6.1 Soil drainage

Usually drainage is accomplished by embedding tiles into the ground around the building site to collect and drain the water in the soil away from the property. If the volume of water is large, as when the foundation or structure extends below the water table or below sea level on property near the ocean, water pumps must be installed. These two methods are limited to porous, sandy soil.

Where the subsoil contains strata of fine silt or clay which must be drained and dried out, more exotic and expensive methods are required, such as electroosmosis.

In the process called electroosmosis the colloidal particles in silt and clay-type soils usually have a negative electrical charge which is used to advantage. These charged particles try to move to a positively charged electrode, as do metallic ions in solution when they are being electroplated. Since the electrically charged soil particles are so large compared to ions, their movement through the viscous soil is negligible. Their associated negative hydrated ions move away (are repelled) to the opposite (negative) electrode. In this way the hitchhiking water molecules are driven out of the soil. Thus the applied voltage pulls positive ions, such as hydrogen or sodium ions and their attached water, into a sump from which the water can be pumped to the surface.

In practice the negative electrodes could be a small steel shell driven down into the clay stratum and attached to the negative lead of a voltage supply. The positive lead would be attached to the nearest metal pile or rod. The accumulated water and the liberated hydrogen gas enter the steel shell through holes in the shell for this purpose. The water would be lifted up through the inside of the shell with a pump assisted by the generated gas pressure. A cluster of negatively charged steel-rod electrodes would surround one steel-shell electrode. Although the electric current is not large, about 50 volts per meter (yard) would be needed between the electrodes of the steel shell and the steel rod to wring water out of the clay stratum electrically.

2.6.2 Compaction methods

Compaction is time-consuming. The rate of settling, rapid at first, slows down after a few days. Unfortunately the compaction decreases rapidly with depth and is effective only a few meters (yards) down from the surface. The settling rate must be carefully monitored. Months may pass before the ground achieves the desired compactness. Careful scheduling is required to avoid an expensive delay in construction. Thus soil treatment methods apply best to the easily reached topsoil, virgin soil, or ground recently used for agriculture.

Improvement of topsoil by surcharge loading is a relatively easy process. There are several ways to compress the soil by temporarily overloading it. Earth-moving equipment can be used to cover the surface of the building site with a heavy layer of stone, gravel, or other available material that can be moved into place. The temporary covering can be moved around every few days to get more uniform compaction. Another method uses an ion-exchange process to convert the clay into the more stable calcium bentonite form. A visually spectacular method uses huge tanks, which can be moved into place and filled with water.

Impact and vibration methods of compaction are more rapid, but they require the use of specialized equipment. In the impact method for surface strata, a very large and heavy hammerlike mass is lifted high above the ground and dropped with an enormous thud. The ground is given time to react to the compression waves and the resulting soil vibration. Then the process is repeated many, many times before moving the equipment to an adjacent area.

The vibration method for compacting deep soil strata consists of resting a substantial steel shaft on the surface and letting the shaft's vibrations go down to the strata being compacted. Considerable power is necessary to transmit the vibrational energy down to the desired strata. This method is capable of reaching greater depths than other methods of compaction. The shearing stresses (layers of soil that slide on each other) generated by the tool promotes subterranean drainage and the collapse of nearby cracks and fissures.

2.6.3 Chemical treatment of clay soils

For weak sodium bentonite clay deposits near enough to the surface to be excavated and mixed with chemicals, conversion to a more stable form of clay is possible. One method uses a salt-coagulation process. This is similar to the way suspended soil will precipitate in muddy rivers to form deltas and sandbars when the river reaches salty ocean water.

For this sodium chloride chemical treatment to be effective, the sodium bentonite clay particles in the soil must have adsorbed insufficient sodium ions to completely neutralize the negative electrical charge on the clay particles. This residual negative charge gives the particles the unwanted, fluidlike thixotropic gel properties. If a little more sodium ions were adsorbed, the clay particles would be neutral and lose their objectionable thixotropic properties. But if the least amount of excess sodium ions are adsorbed, the particles will become positively charged colloidal particles. Obviously, mixing just the right amount of sodium chloride with the soil to produce the desired neutral charge on the particles is tricky. This method seems too risky to be used on a building site unless it is extremely well supervised by an expert.

Another chemical treatment involves mixing the sodium bentonite soil with calcium hydroxide (slaked lime). The high concentration of calcium ions causes an ion-exchange process to replace the sodium ions in the bentonite with calcium ions to produce calcium bentonite. In this way sodium bentonite can be converted to the more stable and better load-bearing calcium bentonite. Note that adding calcium chloride would provide the needed calcium ions, but calcium chloride is a more expensive chemical.

Another method involves mixing cement with the clay topsoil to release the needed calcium ions slowly and to provide some cementitious properties.

Each of these chemical procedures depends on the chemical condition of the soil and would require extensive laboratory testing and supervision.

2.6.4 Piles and caissons

Where there are deep underlying layers of weak, compressible soil, it is necessary to use piles to reach a load-supporting stratum (Fig. 2.4). The piles are required if the subsoil cannot be compacted enough to support a structure. In effect, piles are an extension of the building's foundation. Their purpose is to transmit the working load of the building into the bedrock or other load-bearing stratum. When piles are driven into soils, such as loose sand, compaction occurs from both the vibration of the pile driving and the displacement of the sand to make room for the pile.

Many varieties of load-bearing piles have been developed to use with the different kinds of stratum formations found beneath building sites. Precast reinforced concrete piles are used if the ground has strata of loose soil, such as sand, silt, or clay. Piles depend heavily on the frictional forces of the soil along the surface of the concrete piles to support their loads. Many lengths of piles can be joined together

42   Chapter Two

Figure 2.4 Pilings for building foundation. Pilings driven into unstable coral to reach bedrock are waiting to be cut to ground level for the foundation.

with epoxy cement or by welding the steel reinforcements together. In this way a pile is able to reach the required depth.

Concrete piles are usually made of a number of straight steel bars in the center of a continuous spiral network of steel rods forming a cage, which is encased in concrete. The piles are usually a foot or more in diameter with a square or hexagonal cross section.

A soils engineer has already taken core samples to determine about how deep the piles must go to reach a firm stratum. The actual depth is later confirmed by the resistance encountered.

As the pile is fitted into position on the ground by a derrick, pile and pile driver are lined up carefully so that the pile will be vertical, to a very close tolerance, to avoid going into the ground at a slant. A wood cap is put on the exposed end of the pile to avoid chipping the concrete as the hammer is lifted high and dropped. The energy of the blow on the top of the pile is so great that sometimes the protective wooden pad catches on fire.

Precast piles can be driven into the ground with a screwing motion, by a vibration driver, or by pounding them into the ground with a huge hammer, depending on the ground conditions. The shell can be used over and over to produce more concrete piles.

Any pile with a cylindrical shape can be driven into the ground so that it turns with a screwing motion as it penetrates. This is done by attaching a helical blade or a short screw thread to the end of the pile, causing the pile to rotate as it is driven down into the ground. Also, the screw blades can be incorporated into the reinforced concrete pile when it is made. Their purpose is to provide the pile with additional support from the soil. And by loosening the soil around the pile as it is driven into the ground, much of the friction from the soil is eliminated along with the compressional energy stored in the pile.

Structural steel piles with an H-shaped cross section are used to penetrate hardpan or boulders. Their shape makes the poles more sturdy and causes less soil displacement. These piles may extend down 45.7 meters (150 feet) or more into the ground to reach a supporting stratum, and by welding one end of an H pile to the top of the one already driven into the ground, lengths of 60.9 meters (200 feet) can be used. For weeks these piles are driven into the soil with a pile driver, making the familiar boom-siss-boom sound that can be heard a mile away.

The steel-reinforced concrete piles have a vertical scale on the sides in fractions of an inch, so the distance a pile sinks into the ground with each blow of the driver can be observed. A pile can sink several feet with each blow when passing through top soil. Progress is carefully monitored until the pile sinks only a fraction of an inch at a time, thereby showing that it has reached its intended destination. As predetermined in the site investigation report, the distance the pile sinks indicates entry into different strata until, finally, it reaches the targeted stratum.

Another method produces a formed-in-place concrete pile. This involves threading a sturdy steel mandrel into a thin steel shell-like cylinder. This assembly is driven deep into the ground and additional lengths are attached as needed. The shell assembly is driven into the ground until it reaches the desired depth, as shown by the soil resistance. Then the mandrel is removed and the shell filled with concrete. The external steel shell takes the place of the steel reinforcement normally used in concrete piles.

If additional load-bearing support is needed in especially weak soil strata, a caisson can be poured into boreholes sunk into the ground. These caissons can be as large as 1.5 meters (5 feet) in diameter at the bottom of the pile. A special fitting is attached to the tip of the boring tool to enlarge the hole for the caisson. A reinforced steel cage is inserted in sections and the sections are welded together until the bottom of the hole is reached. Concrete is then poured into the hole around the steel rods to make a formed-in-place reinforced caisson pile, which requires the usual 28 days to be fully cured.

A cluster of 20 or 30 piles may be anchored together and cut down to ground level. The exposed ends are covered with a concrete cap to become part of the foundation. On a major building site in Hawaii, these clusters will practically enclose the perimeter of the building site and will be tied together with sublevel reinforced beams.

2.7 Our Environmental Responsibilities

With the growing awareness that our planet has limited natural resources, long-standing practices of the past are coming under scrutiny. Municipalities are encouraging our formerly "throw-away" society to become aware of the benefits of reusing or recycling almost everything used in our everyday life. This includes paper goods, clothing, furniture, appliances, and construction materials. The contaminated air, water, and ground are serious problems affecting everyone. Urged on by individuals and organized citizen's groups, the government is slowly empowering agencies to survey and monitor our major sources of pollution, including atomic energy plants, many industrial practices, and the armed services.

2.7.1 Recycling used building materials

The recycling of used building materials is becoming necessary because of our diminishing natural resources, the increasing scarcity of dumps and landfills for the disposal of debris, and out of concern for the energy investment these materials represent. The Department of Energy is sponsoring research and developing the technology to make the recycling process efficient and profitable.

To clear a building site for another structure, the demolition of a large building is usually accomplished using explosives. Before demolition, though, it is possible to remove and recycle many of the usable bulk materials, such as wood, aluminum, steel, glass, and concrete. Afterward some of the rubble can be put to other uses. In the future it is conceivable that concrete rubble will be recycled as cement.

The recycling of good, used lumber is just one example of how our natural resources are being conserved. Many contractors reuse, four or five times, the plywood forms for poured concrete structures. Many of the large plywood sheets are reused as partition walls. Short lengths of used wood are cut into wedges to hold the forms in place. These wedges cause less damage to the plywood forms and are easy to remove. Finally, instead of burning scraps of wood that cannot be reused, they are shredded for use as soil compost.

There are growing incentives to extract usable materials from building structures facing demolition. Waste disposal is becoming increasingly difficult and expensive. Landfill sites are overflowing with municipal waste. Strict government restrictions on ground and water contamination threaten to close other sites.

Energy conservation is another consideration, especially for materials with high energy requirements for their preparation. Awareness of harmful environmental effects of burning and incineration are resulting in municipal and EPA restrictions. But is large-scale recycling feasible in the building industry?

In efforts to promote the recycling of solid waste for building materials, the National Bureau of Standards, the EPA, and the Bureau of Mines have prepared feasibility studies on this subject.

Concrete rubble was reused in the bombed regions of Europe after World War II. When the rubble was crushed and ground, it proved to be a practical source of fine and coarse aggregate to make new concrete for rebuilding war-damaged cities. Over the past decade the American Concrete Paving Association and other professional organizations have held symposia on the feasibility and technology of recycling concrete rubble into usable materials for new construction projects. Other materials are recoverable from buildings: steel, aluminum, ceramics, and glass. Once the economics of their reuse are favorable, the technology will follow and the supply and demand will be assured.

2.7.2 Occupational Safety and Health Administration

OSHA was created as a part of the Department of Labor to encourage both employers and employees to prevent on-the-job accidents and health risks due to exposure to hazardous materials used at the workplace. Created in 1970, OSHA had its rights and responsibilities greatly expanded in 1986.

Of special interest is the Material Safety Data Sheet (MSDS), which manufacturers must supply for each product that is purchased by contractors (and other consumers) for use on a building site. The contractors or employers are required to make these MSDSs available to their employees and have each employee sign a statement assuring that the sheets in the MSDS files have been read. An MSDS lists each toxic chemical contained in a product, gives the toxic exposure limits, and supplies information for immediate first-aid treatment if someone is accidentally exposed.

More information about OSHA will be found in later chapters where appropriate.

2.7.3 Environmental Impact Statement

The National Environmental Policy Act of 1969 created the EPA. Its goal is to "create and maintain conditions under which man and nature can exist in productive harmony" with emphasis on air, water, and ground pollution.

Of direct concern to the developer and builder is the requirement to prepare and file a preliminary EIS reporting the need for a structure, or a negative impact statement if there is no anticipated problem. This would be followed by the preparation of a full EIS. Such a statement would contain available data and new measurements, as needed, to support an assessment of the impact the proposed development would have on the air, water, or ground pollution, including additional air pollution from increased traffic created by the use of the proposed building.

The EPA regulations are usually supplemented by state regulations. An EIS can be technical and time-consuming. To avoid building permit delays and possible litigation, the services of a specialist who is competent in this field should be enlisted.

2.7.4 Environmental response, compensation, and liability

CERLA, enacted by Congress in 1980, is a branch of the EPA. With an initial appropriation of $1.6 billion, this was the first federal law established to deal with the dangers posed by the nation's hazardous waste sites. More than 1192 sites are on the ever-expanding list, which keeps growing as more sites are discovered.

CERLA, commonly called the Superfund, becomes involved if the building site has a history as a municipal waste dump or storage facility, an atomic waste storage area, a gasoline service station, an automotive repair shop, or other businesses using or storing toxic materials that could have been spilled or leaked into the underground water supply. Disposal and treatment of soil contaminants are becoming major problems. Therefore, more emphasis on the prevention of contamination can be expected.

Procedures require reporting the situation to CERLA officials, making an inspection to determine the nature and extent of the contamination, and treating or completely removing the soil.

The contaminated soil at building sites is usually confined to, approximately, the top 1.5 meters (5 feet) of soil. There is no universal treatment possible because of the diversity of contaminants. Each kind of contamination requires its own unique chemical or biological treatment.

Remedial options include in-place treatment, involving chemical conversion to a nontoxic or insoluble substance or biological degradation. The commonly used last-resort method is to remove and dispose of the contaminated soil elsewhere at a dump or at a soil reclamation plant. This is still a young technology, and although theoretical procedures abound, their successful applications, beyond experimental pilot studies, have been meager.

Sites of gasoline and auto repair stations contain contaminants such as tetraethyl lead from gasoline, lead sulfate from the electrolyte in lead storage batteries, chromium compounds in antifreeze radiator fluid, and additives used in lubricating oils and greases. For them, choosing an in-place soil treatment may be the best method of reclaiming the soil.

Perhaps the most effective in-place soil treatment is the biodegradation of organic contaminants by bacteria. (Much of this technology has been gained from the study of toxic pesticides used in agriculture.) Many chlorinated organic compounds, such as polychlorinated biphenyls (PCBs) formerly used to fill transformers for electrical insulation, have been degraded by combinations of bacteria and fungi. Since detoxification often involves oxidation, the use of aerobic bacteria is most useful. Certain soil conditions, including soil pH, are important requirements for the bacteria to thrive. One advantage of microbial methods of soil treatment is that the organism multiplies as the process continues.

Chemicals used in the treatment of soils do not replenish themselves as do bacteria. The products of the chemical reactions remain in the soil indefinitely, in some chemical form which would usually be another contaminant.

A method that can work, if it meets the approval of authorities, is to dilute the concentration of the pollutant in the soil by adding clean, pollutant-free soil in order to meet the required standard. Unfortunately this does not reduce the amount of the original pollutant at the site.

Any kind of soil treatment involves enormous bulk and is time-consuming and costly, and some of the products of degradation of organic compounds can be toxic. Disposal of soil contaminants is becoming a major problem. Therefore, emphasis on prevention of contamination can be expected.

Most cities have their own regulations and building codes to ensure compliance with CERCLA. Each municipal building department and department of health is responsible for providing the required information and for compliance.

The CERCLA Superfund for massive cleanup operations of federal sites faces nearly every conceivable kind of contaminant. The research, development, and cleanup are costly, taking years to study the situation and much work to remedy the shortsighted practices of many years' duration. However, the spin-off technology should greatly benefit the future treatment of building sites. For instance, efforts to clean up massive spills from oil tankers and pipelines requiring large-scale research and testing have resulted in the development of new technologies for treating petroleum contamination. As always, prevention is the best policy. There is hope for the future.

2.7.5 Structural environmental hazards

Special environmental hazards should be considered when designing a foundation and a structure. Although building codes provide mandatory protection against chronic situations, it is important in these times of increasing litigation to consider the environmental and climatic trends, especially if the region is subject to earthquakes, high winds, hurricanes, tornadoes, torrential rains, or floods.

As a protection against hurricanes and high winds, special straps should be used to secure foundations to walls and walls to roofs. Because of enormous financial losses, many insurance companies have discontinued writing homeowners' policies for residents in some hurricane-prone areas. Insurance policies in earthquake-prone areas are prohibitively expensive, and therefore many residents do not carry this insurance.

Increasingly, steel, plastic, and other durable materials are being used in home construction because of the year-round continuous destructive activity of corrosion, fungus, termites, carpenter bees, and carpenter ants in the warmer latitudes.

Where termites thrive, a two-day treatment involves covering homes tightly with plastic tents to confine the pumped-in gases that are toxic to termites. (This is a common sight in Hawaii.) When the tents are dismantled, the gases are released to pollute the air and threaten the ozone layer. Within a few years, the termites and other insects return to reinfest the homes and repeat the cycle.

All of us need to be involved in the challenge of protecting our limited natural resources. Cleaning up the pollutants of the past will cost enormous amounts of time, energy, and money. As always, prevention is our best policy.

2.8 Soil as a Growth Medium

For plants to thrive, there must be an optimum amount of moisture, nutrients from the soil and the air, and sunlight. Most of the metallic elements are tied up in minerals, inorganic complexes, and insoluble salts which become available by weathering. The nonmetallic elements, such as nitrogen, are mostly in organic compounds. Microorganisms in the soil make these compounds available to the vegetation. Drainage of water draws air down through the soil to the roots. Humus helps retain moisture in the soil during dry periods. And sunlight is necessary for photosynthesis to take place in the leaves.

2.8.1 The physical nature of soils

Soil is a three-phase system consisting of solids, water, and air. The solid particles in the soil are coated with a thin, invisible film of adsorbed water. Even when particles are air-dried and crumbly to the touch, this film of water remains in place to separate the solid particles. This film forms a network of continuous water paths for soil nutrients to travel through the soil. The extent of drainage will determine the amount of air (the third component) trapped in the soil.

An important part of soil is the organic matter found in humus (composted vegetation), which helps the soil retain moisture and provides nutrients through the action of bacteria breaking down the natural high polymers, such as lignin, cellulose, and proteins. Humus particles usually have large surface areas.

As the size of the soil particles decreases, the surface area per gram increases. However, if the soil is examined more closely, most particles will appear to be porous. This soil porosity adds much additional surface area. In clay soil, the actual microscopic cavities are filled with water. Only by driving out the bound water with heat, in a vacuum, can the true surface area be found. Then the surface area can be measured by finding the amount of nitrogen required to coat these tiny cavities, and many other surfaces, with a monolayer of nitrogen molecules. From the known area occupied by one nitrogen molecule and by calculating the number of molecules used, the true surface area is obtained.

Particles of colloidal-type soil are about one thousandth of a millimeter in diameter. They have a true surface area of more than 1 acre per ounce. Thus whether they are composed of sheets or clumps of atoms, the surface area available for the ionization and adsorption of ions is enormous.

The usual negative charge of soil particles comes from the ionization of positive ions from the surface of these macromolecules. Thus mineral particles lose sodium or potassium ions to acquire a large negative charge. Large-surface-area particles with their exposed negative charges strongly repel one another and do not clump when they collide. Instead, these tiny particles remain dispersed with the fluid-like properties of the colloidal state in the wet soil.

2.8.2 Permeability of soil to water and nutrients

Permeability is a unique problem for some soils because of their porous structure, large surface areas, and the large negative charge of their colloidal particles. It is little wonder that there is a drag on irrigation water and nutrient solutions as they percolate down through the soil, pushing the trapped air ahead of them.

A working model of a colloidal clay soil would be a mass of large-surface-area particles with many exposed negative charges. Their reaction to inorganic fertilizers depends on the charge their ions have in solution. If the nutrient is a potassium or other trace element salt which gives a positively charged ion, adsorption on the negative site prevents their movement down through the soil. Nutrients such as nitrate and phosphate ions with their negative charges are repelled by the negative charge on the solid particle and, therefore, move freely through the soil.

However, the adsorbed ions do not sit placidly on the surface. Because of their thermal energy, these adsorbed ions tug at the attractive force holding them to the surface. This force of attraction between opposite charges acts like a rubber band that allows the adsorbed ion to move back and forth. Some of these metallic ions will manage to free themselves from the surface momentarily. If the particle is surrounded by a high concentration of another ion, one of these may be pulled to the surface as a replacement. This is known as an ion-exchange process. Ion adsorption and ion exchange play vital parts in the storing and movement of ionic nutrients throughout the soil.

Soil color often indicates the quality of drainage, the amount of oxygen in the soil, and the oxidation state of iron. Red or yellow soil shows that a high oxygen content must be present to oxidize iron to ferric oxide. Good drainage results when oxygen from the air is drawn into the soil by irrigation water. A bluish soil indicates ferrous iron and poor soil drainage, producing anaerobic conditions or lack of oxygen. Light gray or white soil usually shows the absence of iron and other minerals due to excessive leaching of metallic salts from sandy soil. A dark-colored soil indicates the presence of much organic matter.

The amount of metallic ions available in solution in the soil also depends on the soil pH. Fortunately, most of these ions are soluble over a range of pH from 5 to 8. Heavy rainfall produces the lowest pH and a more acidic soil. Alkaline soil is usually found in arid regions. Therefore, the type and amount of vegetation that can be grown in an area depends not only on the kind of soil, the soil pH, the use of fertilizer, and the amount of rainfall, but also on the rate at which the nutrients can filter down through the soil to the root systems.

2.8.3 Major soil nutrients

It is customary to refer to the nutrient in a fertilizer by the name of the essential fertilizing element rather than the name of the compound in which it occurs. For example, fertilizer analysis usually reports the percentages of nitrogen (N), phosphorus (P_2O_5), and potassium (K_2O) (the "big three"), which are listed on fertilizer packages as the major soil nutrients needing to be replaced in large amounts. Calcium, magnesium, and sulfur are often included, even though they are usually plentiful in the soil.

The criterion used for selection as an essential element (nutrient) is for the element to be either contained in the plant or known as essential to its metabolism. The list is somewhat arbitrary since it depends on our present state of knowledge. Carbon, hydrogen, and oxygen—abundant in the atmosphere—are not included. Aluminum, cobalt, selenium, and silicon are not included because, as yet, so little is known of their purpose for being in the plant.

A convenient soil model would be a particle with a large surface area and a multitude of chemical adsorption sites. Nutrients compete with one another for these adsorption sites. It is important to keep in mind the potential synergistic effects. The presence of one nutrient can increase or reduce the effect of another. This is in addition to the obvious chemical reactions. For example, there can be a biological synergistic action that is enzymatic in its catalytic effect on plant growth. Thus a balance of nutrients, including major elements and trace elements, may need to be considered.

Since many soils are slightly alkaline, the negatively charged colloidal soil particles readily adsorb the positively charged ammonium cation, whereas the negatively charged nitrogen-containing nitrate ion is repelled and moves relatively freely throughout the soil. These nutrients, adsorbed on the soil particles, are in insoluble compounds or are present in humus and stored in the soil.

Nitrogen in the air is made available for plant use by nitrogen fixation bacteria. A small amount comes from lightning discharges; a trace comes from the smog of engine exhausts. Lightning discharges and internal combustion engines produce oxides of nitrogen and ozone from the high-temperature reactions of the oxygen and nitrogen in the air. Further reaction with petroleum products in the air forms smog.

Nitrogen fertilizers can be introduced into the soil as ammonia gas fed into irrigation water or as a solution of ammonium nitrate or other nitrate salts. Soil bacteria convert ammonium ions to nitrate ions for use by the plant. These bacteria thrive in soil with a pH of between 6.0 and 7.5 and at a temperature between 10 and 32°C (50 and 90°F).

Nitrogen compounds are needed in the soil as part of amino acids occurring in proteins, coenzymes, and chlorophyll. The yellowing of normally green leaves (chlorosis) is usually a sign of nitrogen deficiency. Nitrogen is necessary for the stems, leaves, and flowers of plants, but is not sufficient alone because of the synergistic effect of other nutrients.

The most common inorganic forms of soil phosphorus are the negatively charged mono- and dihydrogen phosphate ions. They are not adsorbed on the negatively charged soil colloids. Phosphates can occur in the soil as insoluble salts of calcium, magnesium, iron, or aluminum. The solubility of these salts depends on the soil pH. Iron phosphate precipitates if the pH is more alkaline than 4. Aluminum phosphate precipitates from the pH of around 6. Calcium phosphate does not precipitate until the soil actually becomes alkaline at a pH of 8 or 9. Some of the phosphorous is also bound up in humus in an organic form.

Phosphorus, needed for root formation and for the development of seeds and the promotion of growth in leafy vegetables, occurs in coenzymes and phospholipids. Phosphorus also holds in check the excessive use of nitrogen by the plant and serves to delay maturity and fruiting.

Potassium ions become available as natural soil nutrients from the slow decomposition of minerals, such as feldspar and potassium clays. However, these minerals are so insoluble that less than 1 percent is available at any one time as potassium ions. The potassium ions not used directly by the plant's root system are stored in the soil by adsorption, especially in the negatively charged colloidal clay soils. Potassium fertilizers contain soluble potassium salts. Potassium, contained in coenzymes, helps regulate the uptake of oxygen, carbon dioxide, and water vapor by the leaves. Insufficient potassium ions in the soil can cause stunted growth, spotted leaves, and weakened stems. This nutrient is necessary for carbohydrate metabolism and cell division.

Calcium, another important nutrient, helps create strong plant membranes and cell walls. For example, the crispness of apples is due to the presence of calcium pectate in the cell wall. Calcium salts, abundant in most soils, are not added as fertilizers.

Magnesium, similar to calcium in its action in the plant, occurs as bivalent positive ions that are tightly held by the negative charge on the colloidal clay particles. Unique because it is a component of chlorophyll, magnesium utilizes the energy of sunlight to promote plant growth. A soil deficiency of magnesium and, thus, chlorophyll results in the yellowing of leaves. The magnesium ion competes with other positively charged ions, such as calcium and potassium, for adsorption sites on soil particles. An excess of either of these positively charged ions can reduce the adsorption of magnesium and produce a nutrient deficiency. This illustrates one way that nutrients added to the soil (or to our diet) are synergistic.

Sulfur, another essential soil nutrient, is needed to a lesser extent by the plant. Contained in some coenzymes, sulfur is needed for the synthesis of chlorophyll, is contained in some proteins, and is important in seed formation and in the growth of vegetables, such as cabbage and turnips. Many sulfur compounds are characterized by a strong, noticeable aroma. Manure provides sulfur for soil enrichment and gets its odor from organic sulfur and nitrogen compounds. Gypsum (calcium sulfate) also provides sulfur for soil enrichment. Unfortunately, the controversial acid rain contains sulfur in the oxidized form (sulfurous acid and sulfuric acid) so harmful to trees, other vegetation, and people.

2.8.4 Trace soil nutrients

The generally accepted six nutrients needed in trace amounts are boron, copper, iron, manganese, molybdenum, and zinc. These trace nutrients are needed in very small amounts and occur in organic compounds such as hormones and enzymes. They are often added to the soil as chelates because, unlike the metallic ions from inorganic salts, chelates are insensitive to changes in pH. Chelates are organic complex salts that are formed by metallic ions, such as calcium, iron, manganese, and zinc. Their advantage is their solubility in water; they are stable in modest swings of pH, and they exist in a form from which the metallic ion is readily available for plant use.

Enzymes, organic compounds containing trace elements, regulate chemical reactions occurring in the plant. Iron, zinc, and manganese occur in coenzymes. A coenzyme serves as an activator of an enzyme.

Hydroponics, the propagation of plants with the roots growing in tanks of a liquid medium, makes it easier to investigate the role played by the trace elements and the major nutrients. Also, much research is being conducted by applying nutrients directly to the leaves. From these small-scale well-controlled ecosystems it is possible to study and improve the nutritional value to the consumer of fruits and vegetables. Most of the concern, however, is given to the quantity and appearance of consumer crops. The following is a brief listing and description of trace elements, showing their value and the need for further basic research.

Boron, occurring in borax and tourmaline, is a vital plant nutrient in trace amounts up to 50 parts per million in soil. Although its part in plant life is not well known, boron has been reported to promote protein and starch synthesis and the transfer of water in the plant. Boron is suggested as being important in root development and for seed and fruit formation. It appears to be involved in nitrogen fixation.

Numerous copper minerals, such as cuprite and malachite, are sources of copper in the soil. Since these are relatively insoluble, the availability of copper ions in the soil is sensitive to the soil pH. The ease with which this multivalent element is converted between its cupric and cuprous oxidation states causes it to act as a catalyst in enzymatic formation of carbohydrate and protein metabolism. It is thought to be involved in nitrogen fixation by plants. Although evidence of copper deficiency is subtle, reduced plant growth and crop yield can result from a lack of copper. Enzymes containing copper are reported to influence the uptake of moisture. Trace amounts can provide resistance to some plant diseases. Copper is used to promote root crops, such as carrots, by the addition of a small amount of copper sulfate to the irrigation water. Copper in the amount of 50 parts per million is recommended in the soil.

Iron, hardly a trace element in the soil since its optimum concentration is given as 250,000 parts per million, is common in many soils as iron oxide, iron carbonate, or other mineral forms. Iron chelates are recommended for addition to the soil over the inorganic iron salts because of their pH stability. A deficiency of the available iron in the soil causes a reduction of photosynthesis, a condition known as chlorosis, indicated by a loss of the green chlorophyll in the leaves. In the absence of iron chelates, chlorosis can occur if the pH of the soil increases above 7, precipitating ferric iron as ferric hydroxide. To remedy this condition, the soil can be made more acid (indicated by a decrease in pH) by the addition of animal manures. The addition of iron sulfate or iron chelates in the irrigation water produces an acid effect on the soil besides providing more iron.

Manganese ion will precipitate as the oxide if the soil is slightly alkaline. Like iron, manganese is a multivalent element essential to form the enzymes which accelerate slow reactions producing chlorophyll. Manganese is essential for the production of seeds. Dwarfing of plant growth indicates inadequate amounts of manganese and iron in the soil. Alkaline soil can cause a deficiency of manganese, resulting in chlorosis. However, if the soil pH is more acid than 5, manganese becomes so soluble that a toxic concentration may result. If additional manganese is needed, it can be added as manganese sulfate, or if its concentration is too high, it can be reduced by alkalizing the soil with lime or calcium carbonate. Manganese at about 2500 parts per million is suggested for most purposes.

Molybdenum, another multivalent element, is reported to be required in trace amounts by nitrogen-fixing bacteria. Usually present in the soil as the negative molybdate ion, only 2 parts per million is needed to form plant proteins and to provide nitrogen fixation and protein synthesis. Mustard, legumes, and cauliflower are reported to be sensitive to the presence of molybdenum in the soil.

The principal source of zinc ion is from the weathering of zinc-containing minerals in the soil. Zinc compounds are far more soluble than iron or manganese. The compounds are less influenced by pH unless the soil is made excessively alkaline with lime. A zinc concentration of 100 parts per million is required for growth hormones and seed production. Zinc is important for the growth of beans, corn, and rice; a deficiency is indicated by dead spots on leaves. Like other trace elements, zinc is considered an essential trace mineral in our diet.

Vegetation, vital to our ecosystem, removes the carbon dioxide from the air and uses it as a plant nutrient, thus reducing the threat of the greenhouse effect. Foliage acts like a body of water, cooling the air as the moisture evaporates from the leaves. The clearing of forested areas, particularly by burning, is extremely harmful to the environment and should be strongly discouraged. Planting trees and other vegetation is beneficial to the survival of our planet.

2.9 Hazards of Pesticide Usage

Pesticides, chemicals used to kill or retard the growth of pests, are used for economical reasons and are designed to have an extended useful life. Consequently many pesticides, such as DDT, banned in 1972, will persist in our food chain almost indefinitely. Our efforts to increase agricultural productivity by the use of pesticides threaten to saturate the land, ocean, and all living organisms.

The magnitude of the problem is appalling. Over 200 million tons of chemical pesticides, representing about 1000 different chemicals, are used in California alone each year. Considering the length of time required to run the costly analytical tests for each of these to determine toxic levels and health hazards, it is no wonder that the EPA is unable to keep up. In many cases, analytical methods and equipment needed to determine the levels of tolerance are not available for use in our testing laboratories.

Pesticides can be toxic to humans in several ways. Not only must pesticides be washed off fruits and vegetables before we eat them, but pesticides can enter the food chain at any level to eventually contaminate all of our food.

From an environmental point of view, cultivated land acts like a body of water. In addition, a planted area is important to the carbon dioxide cycle. Soil organisms are also vital to the plant-animal recycling process.

Improper use of land, by either overgrazing, overirrigating, or overproduction of crops, causes the topsoil to be washed away or blown away in a catastrophic dust-bowl fashion. A change in the source of irrigation water can make a profound change in the soil. For many years the topsoil of Imperial Valley, California, was replenished by the suspended soil brought in as irrigation water from the Colorado River, which had gathered colloidal soil from the neighboring states. Then, to avoid the seasonal fluctuations in water level, a large dam was constructed, resulting in both settling of suspended matter and extension of the lake above the dam into canyons containing beds of salts that increased the salinity of the water. Therefore the Imperial Valley not only lost its topsoil rejuvenation but required extensive drainage to avoid the threat of salinification of the soil.

2.10 Irrigation and Industrial Waste Pollution

Some forms of large-scale environmental contaminations have been well-publicized. Eutrophication of the Great Lakes resulted from the drainage of phosphates and nitrates into the lakes from surrounding agricultural land. The resulting upset in the balance of the ecosystem produced an excess of algae and posed a major threat. Acid rain from industrial effluent has become an international concern between the United States and Canada. The massive spread of radioactive aerosols from the failure of an atomic energy plant in Russia made newspaper headlines around the world. The wholesale use of pesticides, such as the fat-soluble DDT, has killed marine and bird life and has long-range effects on humans.

Perhaps less publicized, but growing in seriousness, are the contamination of farmland by the dumping of industrial wastes in nearby waterways and the pressing need for disposal sites for garbage, sewage, and sludge. Poisoning of the adjacent farmlands with a large variety of trace elements, such as copper, cadmium, magnesium, iron, and phosphorous, can kill vegetation and present a serious threat to humans.

Because of the difficulty and expense to arrive at, and set, toxic levels for each of the ever-increasing variety of potentially toxic chemicals that are introduced into our homes and around the countryside, the EPA is now following a more direct approach. Amounts of annual emissions of pollutants from each industry will be made available to the public for interpretation and action. Will the mere magnitude of the emissions cause corrective action to be taken? Only time will tell. We are part of the environment because we are what we eat and breathe.

2.11 Environmental Considerations

With the growing threat of a blanket of smog covering broad regions of the industrialized world, with pollution of our land and drinking water, with global changes in climate due to the greenhouse effect, and with the enlarging holes in the ozone layer, it is time for everyone to cooperate in saving and restoring our fragile environment. It is a well-established fact that grass and foliage remove carbon dioxide from the atmosphere and act like a body of water to reduce the heat. Replacement with concrete and asphalt, on the other hand, breaks this cycle and creates a hot-air barrier that blocks the normal flow of winds and rains. Since buildings and roads are a necessary part of our civilization, what can be done?

There is much that we can do toward maintaining and improving our civilization. Some examples: Careful landscaping around buildings promotes a restful, healthier, safer, more productive living or workplace. Energy-expensive and resource-draining construction materials are beginning to be replaced or recycled from older buildings. Freon refrigerants are being gradually replaced with other gases less harmful to the ozone layer.

It is estimated that each person creates several pounds of refuse each day. How can this be sorted at the source for recycling to avoid overflowing the city dump sites? Much electric energy is used in seasonal heating and cooling. How can better insulation and heat conservation be achieved? Can drainage and sewage water be reclaimed for irrigation and other practical uses?

How much more can we achieve in our building construction practices by better planning, more recycling, and observing better housekeeping methods? See Chap. 10 for suggestions.

Terms Introduced

Aluvian
Caissons
CERLA Superfund
Clay-mineral structure
Colloidal clay soil
Environmental Impact Statement (EIS)
Environmental Protection Agency (EPA)
Eutrophication
Humus
Major soil nutrients
Material Safety Data Sheet (MSDS)
Nitrogen fertilizers
Occupational Safety and Health Administration
Phosphate fertilizers
Piles
Polychlorinated biphenyls (PCBs)
Potassium fertilizers
Rheology of soils
Soil mechanics
Soil profiles
Soil texture
Thixotropic clay soil
Till
Trace soil nutrients
Weathering

Questions and Problems

2.1. How does a clay-type soil differ from other soils?
2.2. Why is it usually necessary to sink foundations well below the ground level in cold climates?
2.3. What is an EIS and why is it important?
2.4. Why are test borings needed and what information should they provide?
2.5. Describe two practical methods of soil compaction and explain why you think they are the best.
2.6. What types of piles are used and what are the advantages of each?
2.7. Explain how a new building or its landscaping changes the subsoil moisture content.
2.8. Name and give the purpose of four essential nutrients.
2.9. List and describe the need for three trace elements in building site areas.
2.10. Explain how ground-clearing practices in countries of the third world increase the greenhouse effect.
2.11. Describe two situations in which knowing the pH of the soil is important.

2.12. How do plants "fix nitrogen" from the air?
2.13. Name a soil that is colloidal and describe its properties.
2.14. Explain how agricultural chemicals harm the environment.

Suggestions for Further Reading

James Ambrose, *Building Construction: Site and Below-Grade Systems,* Van Nostrand Reinhold, New York, 1991.

D. H. Bache and I. MacAskill, *Vegetation in Civil and Landscape Engineering,* Granada, London, 1984.

G. H. Bolt, *Soil Chemistry,* Elsevier, New York, 1982.

Gene Brooks, *Site Planning: Environment, Process, and Development,* Prentice-Hall, Englewood Cliffs, N.J., 1988.

T. Cairney, *Reclaiming Contaminated Land,* Blackie and Son, London, 1987.

Brian Hicks, "Bioremediation: A Natural Solution," *Pollution Engineering,* p. 30, Jan. 15, 1993.

Morton Newman, *Standard Handbook of Structural Details for Building Construction,* McGraw-Hill, New York, 1993.

Kenneth Noll, *Recovery, Recycle and Reuse of Industrial Wastes,* Lewis Publishers, Chelsea, Mich., 1985.

W. L. Schroeder, *Soils in Construction,* 3d ed., John Wiley & Sons, New York, 1984.

Chapter

3

Cement and Concrete

3.1 Introduction   62
3.2 Cementitious Materials   62
3.3 Formulas of Cement Compounds   62
3.4 Portland Cement   63
    3.4.1 Types of Portland Cement   64
3.5 Hydration, Setting, and Hardening   65
    3.5.1 Setting of Gypsum Plaster   65
    3.5.2 Setting of Lime   66
    3.5.3 Setting and Hardening of Portland Cement   66
3.6 Concrete   67
    3.6.1 Curing Fresh Concrete   68
    3.6.2 Water/Cement Ratio   68
    3.6.3 Porosity   69
3.7 Aggregates of Concrete   69
3.8 Admixtures in Concrete   70
    3.8.1 Water Reducers   70
    3.8.2 Accelerators   72
    3.8.3 Set Retarders   73
    3.8.4 Air-Entrainment Agents   74
3.9 Mineral Admixtures   74
3.10 Reinforced Concrete   74
3.11 Prestressed Concrete   78
3.12 Polymer Concrete   78
3.13 Composites   79
3.14 Environmental Durability   80
    3.14.1 The Corrosive Effect of Salts   81
    3.14.2 Sulfating   82
    3.14.3 Other Destructive Gases   82
3.15 Corrosion of Steel Reinforcement   83
    3.15.1 Passive Steel   83
    3.15.2 Multiple Anodes and Cathodes   83
    3.15.3 Rusting of Rebars in Concrete   84
    3.15.4 Chloride-Induced Corrosion   84
    3.15.5 Corrosion Protection   85
3.16 Building Problems Related to Concrete   85

3.17 Health Hazards	87
Terms Introduced	88
Questions and Problems	88
Suggestions for Further Reading	89

3.1 Introduction

Concrete, used extensively in buildings, is one of the most complicated chemical and physical materials of construction. This chapter explains the technology of combining cement, water, and aggregates to form concrete. A detailed report of the hydration and conversion of gypsum and lime to form crystalline solids helps the reader to appreciate the more complex processes of hydration, setting, and hardening of cement. Changes in the properties of concrete by using admixtures are contrasted. Reviewed are the development and the use of microcomposite fibers to enhance the use of concrete. Common problems such as sulfating, efflorescence, and corrosion are examined. Health hazards are identified.

3.2 Cementitious Materials

A substance that forms a plastic paste when mixed with water, bonds to aggregates, and sets to form a solid material is known as a cementitious material. Common examples are slaked lime, which reacts with carbon dioxide to form calcium carbonate; dehydrated calcium sulfate, which hydrates to form gypsum; and portland cement, from which concrete is made.

3.3 Formulas of Cement Compounds

It is customary, when writing the formulas of solid inorganic compounds, to be satisfied with the simplest formula that shows the relative amounts of the elements present. For example, the formula of calcium silicate is usually written as $CaSiO_3$. However, extensive research in the cement industry has determined more precise formulas and structures for many of the chemicals occurring in cement. Two of the four major compounds occurring in cement are tricalcium silicate and dicalcium silicate. The other two compounds are tricalcium aluminate and tetracalcium aluminoferrite. To provide more information, formulas for the four major compounds and their abbreviations are given in

TABLE 3.1 Major Cement Ingredients

Chemical name	Formula	Symbol
Tricalcium silicate	$3CaO \cdot SiO_2$	C_3S
Dicalcium silicate	$2CaO \cdot SiO_2$	C_2S
Tricalcium aluminate	$3CaO \cdot Al_2O_3$	C_3A
Tetracalcium aluminoferrite	$4CaO \cdot Al_2O_3 \cdot Fe_2O_3$	C_4AF

Table 3.1. The formulas are broken down to show the relative amounts of the basic and acidic oxides that make up the compounds.

Note that the cement symbols in Table 3.1 are abbreviations for oxides, not elements. C stands for lime (calcium oxide), A for alumina (aluminum oxide), F for ferrite (ferric oxide), and S for silicate (silicon dioxide). Calcium oxide occurs in each formula and comprises about 70 percent of each type of cement.

3.4 Portland Cement

A patent for making portland cement from limestone was issued in 1824. It got its name from its resemblance to a building stone found on a small island off the coast of England.

In the preparation of portland cement, raw materials are crushed, mixed, and ground to prepare the desired proportion of lime, silica, alumina, and iron. The finely ground mixture feeds into the top of a sloping, rotating kiln, where the mixture is calcined (sintered) as it descends into the lower, hotter area at a temperature rising gradually from 1427 to 1649°C (2600 to 3000°F). Because ferric oxide has a lower melting point than the other oxides in the clinker, it acts as a flux (promotes fusion).

The clinker is cooled rapidly to preserve the metastable (comparatively stable) compounds and their solid solutions (dispersion of one solid in another), which are made as the clinker is heated. Before the clinker is ground to a powder, about 5 percent gypsum is added to regulate the setting of the cement when it is used to form concrete. Its purpose is to coat the cement particles by interfering with the process of hydration of the cement particles. This retards the setting of the cement for a few hours.

Some of the gypsum is partially dehydrated (calcined) as it is ground with the hot clinker. The rapid hydration and solidification of the calcined gypsum, formed when the cement is later mixed with water, can produce a "false set" of concrete, which disappears on further mixing.

3.4.1 Types of portland cement

There are four main chemical components of cement which are combined in different proportions to make up the five main types of portland cement (Table 3.2). Normal portland cement, or type I, is still the standard cement in use. More specialized portland cements include the following: type II, characterized by a more moderate heat of hydration during setting; type III, a high early-strength cement; type IV, low heat of hydration during setting; and type V, a sulfate-resisting cement for use in areas where high sulfate concentrations occur in soil or water. Specifications for these five portland cements are given by the American Society for Testing and Materials in standard ASTM C 150.

Type I, general-purpose cement. Type I cement is the cement normally used in concrete.

Type II, moderate-heat cement. Type II cement is more specialized. It is used when there is danger of the concrete cracking as the setting concrete cools. Freshly mixed concrete normally expands as it hydrates due to the temperature rise from the heat of hydration of the cement. As the hydration slows and the cement sets, the concrete cools. The resulting contraction produces strains which can cause cracks, especially in the absence of reinforcing steel. Type II cement has an increased amount of C_4AF, which slows down the heat-producing reactions. There is also a small reduction in the proportion of C_3S and C_3A in the cement. These are responsible for early fast setting and much of the heat that is generated. Cracks are caused by the contraction that results on cooling after the cement has set.

Type III, early-strength cement. Type III cement gets its early set and strength from being very finely ground. This provides more surface for reaction with water. Thus the hydration and set processes are accelerated. Also, since C_3S hydrates more rapidly than the other cement compounds, its proportion in the cement is maximized.

TABLE 3.2 Composition of Portland Cement*

Type	Use	C_3S	C_2S	C_3A	C_4AF
I	General purpose	50	23	11	8
II	Moderate heat	46	29	5	13
III	Early strength	57	17	10	8
IV	Low heat	27	49	4	12
V	Sulfate resisting	38	43	4	9

*In percent.

Type IV, low-heat cement. Like most spontaneous chemical reactions, hydration of cement is exothermic (evolves heat). Since concrete is a poor conductor of heat, the heat of hydration raises the temperature and poses the danger of the concrete cracking on cooling. An alternative is to slow down the reaction and give the heat more time to escape from the cement. Type IV cement minimizes the proportion of C_3S and C_2A, the cement compounds that produce heat of reaction most rapidly. In massive concrete structural units even this slower rate of evolution of heat cannot be tolerated; therefore other methods are necessary. These methods involve using ice in place of water to chill the concrete mix and circulating cold water through pipes embedded in the concrete during setting.

Type V, sulfate-resisting cement. In the reaction of sulfate ion with the calcium in C_3S and C_2S (the high-calcium-containing cement), hydration products form calcium sulfate, which disrupts the structure of the cement.

Portland cement may also contain air-entrainment agents and pozzolan cement. These are indicated by separate letters added to the types of cement listed. For instance, types IA, IIA, and IIIA indicate an added air-entraining cement; and the addition of the letter P, as in type IP cement, indicates type I cement containing some pozzolan cement.

3.5 Hydration, Setting, and Hardening

The action of water on cement can be better understood by starting with the action of water on plaster of paris, and on slaked lime as used in mortar and in whitewash.

3.5.1 Setting of gypsum plaster

Setting of partially dehydrated calcium sulfate (plaster of paris) illustrates the setting and hardening by recrystallization. Gypsum (hydrated calcium sulfate) is a dihydrate having the formula $CaSO_4 \cdot 2H_2O$. When heated to about 200°C (392°F), it loses its water of crystallization to form the hemihydrate (half-molecule of water). The formula is $CaSO_4 \cdot 5H_2O$.

When water is mixed with the hemihydrate, some of it dissolves, hydrates to form the dihydrate, and precipitates from the supersaturated (oversaturated) solution as tiny crystals. If only enough water is used to form a plastic paste (as in forming plaster), a dense, confused mass of needlelike crystals is slowly formed, which interlock to make a strong, hardened solid. This process is essentially the action of water to reform the dihydrate.

3.5.2 Setting of lime

Lime is a general term used in industry to indicate either quicklime (calcium oxide) or slaked (hydrated) lime (calcium hydroxide). Slaked lime is used in stucco, mortar, and whitewash. Quicklime is seldom used as it reacts vigorously with water (evolving heat as it hydrates), is dangerous near organic matter, and is a strong irritant.

The initial setting of a mixture of lime and water takes place in two steps. First, the slaked lime is mixed with water to form a thick whitewash solution. This mixture forms a supersaturated solution which precipitates a weak solid of amorphous (noncrystalline) calcium hydroxide particles. The whitewash bonds to wood or other surfaces to which it is applied. In the second step there is an initial reaction of carbon dioxide from the air with the calcium hydroxide to form crystalline calcium carbonate. The interlacing calcium carbonate crystals produce a strong whitewash coating. Complete conversion of the solid calcium hydroxide below the surface is much slower. This is due to the slow penetration of carbon dioxide through the surface layer of calcium carbonate and to the absence of moisture to promote the reaction. This process is essentially one of supersaturation, precipitation of amorphous calcium hydroxide, and a final slow reaction to form entangled calcium carbonate crystals.

3.5.3 Setting and hardening of portland cement

The setting and hardening of portland cement is much more complicated. This is due to the different proportions of the four main cementitious compounds in the various kinds of portland cement. Each of these compounds may occur in several different crystalline structures. Many of the hydration products are metastable and slowly change from one form to another. Some of the products of the cement hydration even react with one another. Consequently, only a general description of the setting and hardening process will be given here.

When the cement particles are mixed with water, a series of changes occur:

1. Water reacts with the surface of the cement particle, and the product forms a supersaturated solution from which a gel-like mass of fibrous crystals precipitates. This gelatinous coating around the par-

ticle acts as a barrier to seal off the particle from further reaction. However, "free" water slowly diffuses into the gel by osmosis (spontaneous dilution of the gel solution). This water reaches the unreacted surface of the particle and forms more hydrates. The gel swells as this process continues. Finally, the swollen gel ruptures and fills in the spaces between cement particles and aggregates to form a semisolid gel. This network of tiny crystals produces the initial set. The process takes about an hour for standard cement.

2. The metastable crystals slowly interact and recrystallize into larger fibers to form a strong network that characterizes the hardened cement. This "final set," marking the beginning of the hardening process, usually starts about 10 hours after mixing with water. For example, C_3S reacts with water to produce compressive strength more rapidly than the other cement compounds. Its overall reaction with water is given by the equation

$$2(3CaO \cdot SiO_2) + 6H_2O = 3CaO \cdot 2SiO_2 \cdot 3H_2O + 3Ca(OH)_2$$

There are at least two intermediate hydration products in this reaction with water. This reaction only shows the final, more stable products; it does not show the complexity of the process.

The rate of hydration and other important properties of these cement compounds differ widely. For example, C_3S achieves about half of its final strength in 7 days, whereas C_2S takes 3 months. Both reach their maximum strength in about a year. The rate of hydration of cement is dependent on the fineness of grind of the clinker. The finer the cement particles, the more surface area there is for reaction with water. Therefore, the finer the cement is ground, the quicker the setting and stiffening occur. In practice, the hydration of the cement particle only penetrates about a fraction of the way into the surface. Experiments have shown that hardened concrete can be reground and used in place of fresh cement a second time, and even a third time, before the interior of the cement particle becomes completely hydrated. Could reusing old concrete be a solution to the enormous amounts destined for our overflowing municipal dumps? Recycling, discussed in another chapter, could help save our natural resources and energy.

3.6 Concrete

Concrete consists of cement, water, sand, rock, and sand aggregates and admixtures. Admixtures are added during the mixing of the concrete to produce special properties. They can alter the setting, hardening, strength, and durability of the concrete. Some of the common admixtures provide water reduction and air entrainment. The composition of the standard types of cement are shown in percentages in Table 3.2.

3.6.1 Curing fresh concrete

Curing the freshly placed concrete is an important factor in the strength and durability of concrete. Setting, hydration, and crystal-growth sequence should not be interrupted, particularly during the first 48 hours for type III early-setting portland cement. This is twice as long as required for general-purpose type I. The surface of the cement should never be allowed to dry, for this indicates a scarcity of water for the hydration process. If the surface becomes dry, water for the hydration process is insufficient and a serious decrease in the quality of the concrete will result. Frequent sprinkling may be needed. In hot, dry weather it may be necessary to leave the forms in place and cover exposed surfaces with a sheet of polyethylene or other suitable material.

As a general rule, the setting of cement takes twice as long for each drop in temperature of 10°C (18°F). This rule depends on the heat of reaction, for many exothermic reactions are about the same. In practice, type III cement takes about 2 days to cure at 21°C (70°F), but it takes 5 days if the temperature drops to 10°C (50°F). If the air temperature drops to 5°C (40°F) or lower, the water and the aggregate should be heated to keep the hydration and setting processes to a reasonable length of time. If the surface of the concrete freezes before it is fully set, the crystalline structure will be disrupted and a serious loss in strength will occur.

3.6.2 Water/cement ratio

Water is a chemical reactant in the hydration of C_3S for rapid hydration, set, and early strength. The relative amount of water and cement, measured by the water/cement ratio (w/c ratio), largely determines the strength of the concrete. Either too little or too much water weakens the strength of the concrete. Additional water is required to wet the aggregate so it will not cling together. More water, needed to provide fluidity, allows the concrete to flow into a form and to completely fill all spaces around the steel reinforcements.

The smaller the water/cement ratio (increased cement) the higher the compressive strength of the concrete. Yet cement is the most costly component of concrete. The specifications of the individual circumstances and performance needed are often given rather than the precise composition of the concrete.

Impurities in water can affect the strength of the concrete and its setting time, cause sulfate deterioration and efflorescence, and promote corrosion of reinforcing steel. A simple rule is that if the water is potable (drinkable), it is suitable for concrete. However, if there is the least doubt, a sample of the water should be sent to a laboratory for chemical analysis. If city water is to be used, its analysis should be available from the municipal water department.

3.6.3 Porosity

Porosity of concrete leads to both physical and chemical deterioration. Because of the heterogeneous structure of a concrete mixture, some porosity is to be expected. Aggregates can be a major source of porosity if their size distribution is not uniform. If their size exceeds a certain limit, the particles are more apt to settle and clump together in the bottom of the mix. Air bubbles produced during mixing may be purposely replaced by the smaller more uniform distribution of air bubbles by air-entrainment agents. These air spaces, with their passages and voids, contribute to the porosity of the concrete.

If the aggregates are not distributed uniformly in size, the spaces between them leave voids that will fill with water and air. This will require excess cement paste to fill the voids in order to cement the particles together and maintain the strength of the concrete. For a better fit and optimum concrete strength, the ideal shape of the aggregates is cubical, flat, or elongated with a rough surface for good adhesion. They should not be rounded or smooth.

3.7 Aggregates of Concrete

Aggregates make up about 75 percent of concrete. Although thought of as an inexpensive filler, aggregates provide strength to the concrete since they are usually stronger than the cement holding them together. The strength of concrete is lowered if crushed rock is used. Jagged, irregularly shaped particles of rock prevent close packing of the aggregate, thereby requiring more cement to fill the crevices. For best results, particles should be graded in size so the aggregates can fit closely together to form a strong, tightly packed structure.

A thin layer of cement provides fewer chances of flaws and adhesive failure. The tendency of larger rocks to leave voids, increasing the porosity of the concrete, limits their use. ASTM C 33 gives the specifications for aggregates for portland cement. The specifications for lightweight aggregates are given in ASTM C 330. Lightweight aggregates reduce strength, but can provide fire protection and sound and heat insulation.

3.8 Admixtures in Concrete

Admixtures are chemicals added to concrete to modify the physical properties. An admixture often affects more than one property, so side effects must be considered if they are used. For example, water-reducing agents increase workability and can act as set retarders. Water reducers can also increase the early strength of concrete. An admixture has numerous other effects, making it appropriate to group the many effects under a few unifying actions.

A well-established method of measuring the workability and set of concrete is the slump test described in ASTM C 143. This slump test for portland cement provides a standard method to measure the fluidity of a portland cement mix. In the procedure, a metal mold is filled in three layers with the concrete mix. Each layer is tamped with a special rod. The slump of the mix, in height, is measured immediately after the mold is gently lifted away. The metal mold, which is 30.5 centimeters (12 inches) high, is a frustum (truncated cone) with a base of 20 centimeters (8 inches) in diameter and a top of 10 centimeters (4 inches) in diameter. The amount of slump is especially useful in comparing the effect of admixtures on the set of concrete.

Before an admixture is decided upon, it is important to consider alternatives that might achieve the same desired properties. Determine whether there are any undesirable side effects in using the admixture. Does the admixture meet the ASTM C 484 specification for admixtures and the municipal building code? Or, if flowing concrete is used, it should meet the ASTM C 1017 specifications for chemical admixtures. How does the admixture affect strength, workability, durability, shrinkage, and creep? The choice depends on the type of usage, the environmental exposure, and the performance required. If none of the available cements can meet the needs of a project, modification with admixtures is needed. Specifications for chemical admixtures for concrete are given in ASTM C 494. Admixtures can be grouped into water reducers, air-entraining agents, and specialty agents.

3.8.1 Water reducers

A water-reducing admixture reduces the amount of water needed in concrete and improves its workability by acting as a plasticizer. It also increases its early strength. An admixture can reduce the amount of water needed by 10 to 15 percent and thus reduce the amount of cement needed to achieve the same performance. In this sense, a water reducer can be a cement saver.

Excess water is usually added to concrete above that needed for hydration of the cement. This is to provide water for workability. By using a water reducer, extra workability is obtained. Although the primary action of the water reducer is to increase workability, it also increases the strength of the cement. Therefore a water reducer allows reduction in the amount of water usually needed for workability. But the amount of costly cement must also be reduced to retain the same critical water/cement ratio. Thus a water reducer allows a reduction in water, cement, and the cost of the concrete without sacrificing the strength of the concrete.

The chief water-reducing agents are lignophosphate, hydroxy carboxylic acid, hydroxylated polymers, formaldehyde naphthalene sulfonate, and formaldehyde melamine sulfonate. Lignosulfonate also has some of the properties of an air-entrainment agent, being an impure cross-linked high polymer containing several different sugars. It is a sodium or calcium salt of a sulfuric acid derivative and a by-product of the paper industry.

Water-reducing and retarding agents. An admixture added as a water-reducing agent may also have other modifying properties. To illustrate, lignosulfonate, primarily a water reducer, has set-retarding properties as well. Other chemicals that have a set-retarding effect of from 1 to 3 hours, as well as their water-reducing effect, are (1) salts of aliphatic hydroxycarbolic acids, (2) salts of lignosulfonic acids, and (3) polysaccharides (sugars). Sodium, calcium, or triethanolamine salts are used to increase the solubility of the organic acids in water. Sodium lignosulfonate is an exception with only a marginal retarding effect.

High-range superplasticizer water reducers. The superplasticizer admixtures are also known as superfluidizers, super water reducers, and high-range water reducers. Their action is that of a surface-tension-reducing agent (surfactant) that breaks up cement aggregates into smaller groups of suspended cement particles to make the cement mixture more fluid. This class of admixtures has been called superplasticizers because they increase the plasticity and workability of concrete mixes. This increased fluidity can be utilized in three ways:

1. Extra-strength concrete can be obtained only by reducing the water content, thus increasing the relative proportion of cement and the strength of the concrete. This reduction in the water/cement ratio

would normally make the concrete unworkable, except for the presence of the superplasticizer. The water/cement ratio and the water content, since they are proportional, can be reduced by as much as 30 percent, which is a two or three times greater reduction of water than is possible with normal water reducers.

2. A saving in cement can be made by reducing both the water and the cement as long as the water/cement ratio is unchanged. The extra workability provided by the superplasticizer makes up for loss from the reduction in water. Even if savings in the cost of cement may be offset by the cost of the superplasticizer, there is still a saving of energy since the energy needed to produce the cement is usually greater than that required for the superplasticizer.

3. A superflowing, or "flowing," concrete results if no change in the formulation is made. The added fluidity provided by the superplasticizer causes the concrete to be a self-leveling, self-compacting concrete that flows easily around hard-to-reach reinforcing steel.

Superplasticizers are usually classified into four major groups of compounds. They are melamine formaldehyde condensate (one type of polymerization), sulfonated naphthalene formaldehyde condensate, modified lignosulfonates, and sulfonic esters. All of these are organic compounds that are sulfonated as sodium, calcium, or ammonium salts for the convenience of using them as water solutions.

3.8.2 Accelerators

In cold weather, accelerating admixtures help to restore more normal setting and early strength times. Accelerators compensate for the reduction in ambient temperature and the resulting slower rate of reaction. They can be useful in reducing the time required to complete non-load-bearing concrete sections such as walls and floors. Early strength and set development do not ensure greater final strength. Other desirable properties may be reduced.

Many chemicals have been used as accelerators, including calcium chloride, calcium formate, other calcium salts, sodium hydroxide, and sodium and potassium nitrates. The most common accelerating agent is calcium chloride. The usual effect is acceleration of the hydration of C_2S and C_3S, which provides most of the strength of the cement in concrete. Since 1873 calcium chloride has been the most widely used accelerator. However, the use is controversial due to the adverse side effects of corrosion of steel reinforcements and deterioration of concrete. Alternatives are continually being sought.

Over a period of many years, extensive research on the mechanism and use of accelerators has been reported. Both the calcium and the chloride ions have been shown to contribute to its effectiveness. Calcium chloride, when used in amounts of 1 to 2 percent with C_3S and C_2S, accelerates both the rate of hydration and the achievement of early strength.

There are several possible explanations for the accelerating action of calcium chloride. One is the adsorption of chloride ions by the C_3S and C_2S hydrates, influencing the transition from one metastable hydrate to another. A second mechanism proposes that chloride adsorption produces smaller, more numerous hydrate crystals to provide additional nuclei for the precipitation of calcium silicate hydrates, thus hastening early setting and strength. A third mechanism is that the calcium ion of calcium chloride forms a solid solution with the calcium hydroxide formed in the hydration and, therefore, promotes more rapid hydration. A fourth is that the more acid pH observed following the addition of calcium chloride indicates less calcium hydroxide in solution. Lower calcium hydroxide would hasten the hydration of the calcium silicates and, thus, the setting and hardening processes it initiates. Fifth is the catalytic effect of calcium chloride on the hydration of the calcium silicates and, especially, tricalcium aluminate. Possibly, each of these mechanisms is responsible for the accelerating effect of calcium chloride.

3.8.3 Set retarders

A set retarder is an admixture that extends the workability and setting period of concrete. When working in hot climates, set retarders can compensate for the rapid setting due to the increase in temperature, but they are effective only during the first week of setting. The slump test described in the standardized test method for "slump" of portland cement concrete, ASTM C 143, is important in comparing the setting time of concrete. Note that water reducers usually accelerate the setting of concrete. Due to the lower rate of hydration during the early setting, water reducers also extend the setting time.

Accelerators provide early strength for concrete. Water reducers and retarders contain similar ingredients and produce similar results. Organic compounds are often used, such as sodium, calcium, or amino salts of lignosulfonic acid. A second type consists of hydroxycarbolic acid salts. A third type is the carbohydrates. The most widely used set retarder is lignosulfonate, which is one-third carbohydrates.

3.8.4 Air-entrainment agents

As often occurs in science, the discovery of air-entrainment agents was accidental. Beef fat containing glycerol stearate had been used to lubricate the clinker-grinding equipment. Concrete prepared in this equipment had improved durability. The improvement was traced to the air-entrainment action of the beef fat. Entrainment agents, such as the surfactants, act as emulsifiers and foaming agents to improve the plasticity and workability of the water paste. They also reduce bleeding by stabilizing the gelatinous mixture. Other uses for surfactants are washing compounds, soaps, and wetting agents in industry and households.

Air-entrainment additives must form stable air bubbles, and thus are detergents with foaming properties. Most commonly used are abietic acid, fatty acids, and alkyl anionic detergents. Abietic acid has a carboxylated ring structure. Oleic acid (an 18-carbon-atom unsaturated acid) is the usual fatty acid used. The rheological effect of air-entrainment agents is found not only in the cement and its hydration, but in the mineral aggregates of the concrete.

3.9 Mineral Admixtures

Natural mineral admixtures include volcanic glasses and tufts, clay and shales, and diatomaceous earth. Industrial by-products include steel-furnace ash, slag, and rock residues. These materials should meet the specifications for mineral aggregates, fly ash, and pozzolan given in ASTM C 618.

3.10 Reinforced Concrete

Concrete can be reinforced with steel bars (rebars) and with steel cables in prestressed concrete. Figures 3.1 to 3.6 show a few of the varied structures that are possible with this technology. (Steel reinforcements, subject to corrosion, are discussed later in this chapter.) In an effort to eliminate corrosion, the rebars can be galvanized with zinc, but at considerable expense and with questionable results. Copper rebars have been tried without much success.

Rebars have lugs (deformations) at regular intervals to increase mechanical adhesion. The rod size, the lug size and spacing, and the mechanical properties are specified in ASTM A 706, A 715, A 716, and A 717.

Three conditions are usually necessary to promote the corrosion of steel: moisture, oxygen, and an electrolyte (such as sodium chloride).

Cement and Concrete   75

Figure 3.1 Modern concrete construction. Note prefabricated concrete stepped for bleachers, prestressed forms poured, lifted, and interlocked, and utility conduits.

Figure 3.2 Carpenters setting up forms for columns.

Figure 3.3 Reinforced columns under construction, showing formwork at various stages of completion, scaffolding between beams, and supports between floors.

Figure 3.4 Supports for precast floors include scaffolding holding up formwork for columns supporting precast concrete floors. Removable sections to be pulled out by crane.

Figure 3.5 Raw concrete after removal of forms. (*Courtesy of University of Hawaii, UH Relations.*)

Figure 3.6 Concrete structure near completion, with installation of aluminum frame window components. Landscaping has begun.

It was once believed that concrete could protect steel from corrosion by keeping out moisture and oxygen. Instead, corrosion is retarded by the passivity of steel due to the alkalinity of the concrete.

3.11 Prestressed Concrete

Prestressing concrete is a way to compensate for concrete's low tensile strength. A beam is prestressed by stretching a high-tensile-strength cable down the length of a concrete form. Then the cable is put under tension and stretched by using jacks at either end. High-strength concrete is poured into the form around the cable. The concrete bonds to the cable as it hardens. When the jacks are removed, both the cable and the concrete beam remain under the amount of tension desired. This prestressed concrete beam, or structure, is installed in place in such a way that the applied load will have to overcome the stored stress before any torsional load can be experienced by the beam. This is known as a pretensioned beam.

Another method of forming prestressed concrete is to posttension the cable. This requires the cable to be encased in a thin steel or paper tube (within the form) and anchored at the ends before the concrete fills the form. After the concrete has hardened, the cable can be more easily put under tension with jacks at the ends of the beam or structure. This can be done after the beam is in place on the work site. Finally the tube is filled with grout to complete the prestressed concrete beam.

To help get a feeling for how this works, visualize lifting a row of children's building blocks (in a unit) by firmly pressing the row of blocks together. The greater the force used to push the blocks together, the more blocks can be lifted at one time. In this illustration the blocks are being pushed together instead of being pulled together. This engineering discovery has made it possible to design a much lighter structural beam or slab, resulting in the savings of nearly half the usual amount of concrete and an even larger amount of reinforcing steel.

3.12 Polymer Concrete

Polymer concrete (portland cement replaced by a polymer) has a lower rate of water absorption, higher resistance to cycles of freezing and thawing, better resistance to chemicals, greater strength, and excellent adhesion qualities compared to most other building materials.

The most commonly used resins (polyesters and acrylics) are mixed with the aggregate as a monomer with a cross-linking agent (a hardener) and a catalyst to reach full polymerization. Polymer concrete is usually reinforced with metal fibers, glass fibers, or mats of glass fiber.

Polymer-impregnated concrete (PIC) is cured portland cement concrete that is impregnated with a monomer using pressure or a vacuum process. The monomer (most often an acrylic) is polymerized by a catalyst, heat, or ultraviolet radiation. A continuous surface layer is formed which waterproofs, strengthens, and fills the voids.

3.13 Composites

Today composites are finding applications in all types of building materials. In broad terms, composites are mixtures of two or more substances which remain as separate solids in the final product. Mortar and concrete composites have been tested for improved workability, crack resistance, impact resistance, impermeability, toughness, strength, and durability.

Composites can be classified as macrocomposites and microcomposites, depending on the size of the reinforcing material. Macrocomposites occur in a variety of common forms such as laminated-wood veneer, composition board, and woven fabrics of natural and synthetic fibers. Typical microcomposites consist of thin, short fibers suspended in a binder framework (matrix) of another material. Common examples of microcomposites are glass-fiber-reinforced plastic, asbestos-reinforced roofing felt and tar, and fiber-reinforced cement. It is the microcomposite of polymers in concrete that is of special interest in this chapter.

For decades, asbestos has been used in a cement matrix to increase the toughness and strength of cement. Asbestos fibers are hydrophilic (water-liking). Wet cement clings to the fibers with good adhesion. Asbestos cement composites were manufactured as asbestos board, tile, and pipes, and for sound and heat insulation. Or the composite could be mixed on the work site. One such use was to form insulation for steam pipes and other fittings deep in the hulls of ships in shipyards. Asbestos portland cement proved to be a perfect composite. It was cost-effective and appeared to be ideal for cement composites in a multitude of applications.

Unfortunately, following continual exposure to asbestos fibers, workers began coming down with asbestosis, a chronic lung inflammation caused by inhaling the very fine asbestos fibers floating in the air. Some workers developed a dreaded form of cancer that was traced to asbestos.

To complicate matters even further, the world supply of asbestos began to run out. Asbestos seemed indispensable, but what could be found to take its place? The search for a substitute has involved years of intensive research, indicating that asbestos is unique. One of the most promising and cost-effective substitutes resulted from a combination of the technologies of concrete and polymer science. The use of polymer fibers (such as polypropylene) as a replacement for asbestos in cement has received much attention.

The performance of a fiber-reinforced cement composite depends on several factors. The fiber should be low in cost in keeping with the low cost of cement and concrete. Adding the composite should result in a marked improvement over cement alone. The adhesive bond formed between the fiber and the cement must be strong. The size, shape, and orientation of the fiber are also important considerations.

Test trials showed that the chemical adhesion of polypropylene fibers was inferior to that of asbestos, with the failure occurring at the surface of the fibers. Hoping to obtain mechanical adhesion from the mechanical entanglement of cloth in the cement matrix, a woven polypropylene was tried. This resulted in better adhesion along with good impact resistance, but the polypropylene cloth was difficult to disperse in the cement matrix. Tests using fabricated (ribbon-shaped) polyethylene fibers showed improved adhesion, but the fibers were difficult to disperse. Further development has shown that polymer fibers may have their place in the construction industry. These fibers can be mixed with concrete to replace the steel reinforcing bars when load-bearing properties are not required. Polymer-fiber-reinforced concrete (used in curtain walls and concrete floors) provides thinner and lighter-weight walls. Composite fibers are more expensive than steel and glass fibers, and their use is restricted largely to particularly demanding applications. Extensive on-site trials could be required before general adoption.

On the other hand, fibers such as steel and glass provide concrete with strength and stiffness. These fibers are hydrophilic enough to form strong bonds with cement. Steel fibers, used in 0.5 to 1.0 percent by volume in reinforced concrete beams, showed promising results in extensive laboratory tests. Improvement was found in load-deflection and torque-rotation resistance. Crack resistance nearly doubled.

Research shows that a new admixture of concrete incorporating tiny carbon fibers increases its strength against fractures and its resistance to earthquakes.

3.14 Environmental Durability

Although concrete is one of the most durable construction materials, all concrete ages due to the action of the environment. High-grade concrete has a ringing sound when hit with a hammer. Some concrete mixes have had to compromise their durability for early strength and other special properties. Fractures break directly through the aggregate and cement structures instead of separating at the boundaries of the aggregates.

Crushed rock and sand, held together by cement and enmeshed in hydrated cement crystals, provide a multitude of fissures, capillaries, and entrapped air. Through these spaces, moisture and reactive gases can penetrate the concrete. Exposure to the environment causes excessive expansion from changes in temperature and moisture. Air, water, and soil provide sulfates which further deteriorate the concrete.

Although concrete normally can change in moisture content in wet and dry weather, this change is not serious. The presence of a small amount of the hydroscopic calcium chloride can cause destructive swelling of concrete. In addition to expansion of concrete due to excessively high summer temperatures, the drop in temperature in the winter can cause freeze-thaw cycles that ultimately can cause cracks and shifts in building foundations. In Alaska, or in other very cold locations, the heat from improperly insulated buildings can cause melting of the permafrost and collapse of the structure.

The ever-present carbon dioxide in the air dissolves in fog, rain, and snow to produce a weak acid solution that gives concrete a rough surface and sandy texture. Other reactive gases in volcanic "vog" and industrial smog are sulfur dioxide and oxides of nitrogen. Calcium chloride and salt used as deicing compounds are carried onto parking decks and are blown with dust and dirt over the surfaces of concrete buildings. Ocean spray carries this salt along with many other compounds. In this way, the environment joins forces to attack our concrete structures.

3.14.1 The corrosive effect of salts

Rusting of the rebars in reinforced concrete can cause spalling, cracking, and deterioration. This is especially important where sodium chloride and calcium chloride are used to reduce the winter icing of streets and in areas near the ocean where fog and spray carry salt crystals inland. Prevention includes careful washing of aggregates mined around seashores. A protective epoxy coating on the steel rebars prevents corrosion. However, special attention must be given to preventing cracks in the coating. The high corrosion density and attack can become locally intense. The added protection of galvanizing the steel with zinc could be worth the added expense in certain critical applications. Cracks in the zinc coating would not be of any consequence since the zinc coating acts as an electrochemical aid.

3.14.2 Sulfating

Sodium, calcium, and magnesium sulfate can attack cured cement. The sulfate ion from these salts reacts with calcium ion from the hydrolysis of cement and with the hydrated calcium aluminum silicate. The products of the reaction are insoluble calcium sulfate (gypsum) and calcium aluminosulfate. The reaction occurs because these products are more insoluble than the reactants and occur throughout the hardening process. Because the volume of the calcium sulfate and calcium aluminosulfate crystals is larger than that of the solids they replace, the concrete is disrupted and disintegrated by the irresistible force of the expansion. A common source of sulfate salts is the soil around the concrete foundation. Sulfate salts can come from the water used in the concrete and from the aggregates if unwashed aggregates are obtained from areas near the ocean. These aggregates should be washed very carefully.

Protection against sulfates, up to 2 percent in soils, can be obtained by using about one-third pozzolanic cement. As an alternative, type V cement can be used.

3.14.3 Other destructive gases

Pure water and rainwater contain carbon dioxide that is always present in the air. This is not a problem when using our drinking water because there is a slightly alkaline reaction from the many dissolved salts present in trace amounts. Carbon dioxide is usually left in distilled water, giving it an acid pH. This carbon dioxide attacks the calcium hydroxide formed by the action of water on cement (hydrolysis). Any fissures or cracks in the concrete can leach out the calcium hydroxide normally present in concrete. In some cases the carbon dioxide in the water can convert calcium carbonate in the concrete into soluble calcium bicarbonate and cause further extensive disintegration and crumbling. Good drainage of surface water, proper curing, ample reinforcement to avoid crack formation, and smooth, dense concrete are some of the preventive measures that can be taken.

Acid rain, produced by the action of moisture and sunlight, is catalyzed by the dust and fine ash particles that accompany and adsorb these gases. The oxygen and moisture in the air convert sulfur dioxide and nitrogen oxides into nitric acid and sulfuric acid. These acids are very corrosive and, even in small amounts in the air, dissolve the surface of marble, limestone, and concrete. These air pollutants can be carried over hundreds of miles. Prevention is the best solution.

The poor performance of concrete may be caused by the use of inferior materials, by the environmental action of weathering and the chemicals in the air and ground, and by extreme changes in temperature, including freeze-thaw cycles.

3.15 Corrosion of Steel Reinforcement

Steel is embedded in concrete as rebars and as cables in prestressed concrete. Corrosion of this steel can be a serious problem in buildings, bridges, dams, and highways. Where highways and roads are involved, the practice of treating ice-covered roads with salt greatly increases the corrosion hazard.

Several conditions are necessary to promote the corrosion of steel rebars or cables when they are in concrete. The passive coating of ferrous oxide on the steel surfaces must be destroyed. The cathodes must have adequate oxygen to maintain their polarization voltage. Electrons released at the anodes (from iron atoms forming ferrous ions) must find an ionic-conducting path through the concrete to the other electrode. Finally, the conditions must include the formation of a multitude of small anodes and cathodes on the steel surface.

Ideally, for high-quality concrete with a relative humidity of less than 50 percent and in the absence of chlorides, there exists little threat to structural steel. Unfortunately, in practice these ideal conditions are not always realized.

3.15.1 Passive steel

Steel is usually passive to corrosion in fresh concrete with its alkaline pH of about 13. This concrete alkalinity results from the reaction of its cement with water to form calcium hydroxide and some sodium and potassium hydroxides. In this alkaline environment, steel forms a thin, adherent, corrosion-resistant hydrated ferrous oxide coating.

3.15.2 Multiple anodes and cathodes

In reality, a steel rebar embedded in concrete differs markedly over its surface in the amount of chemicals present and in points of stored mechanical strains. These random surface conditions differ enough from one another to form a multitude of concentration cells over the steel surface. The anodes are potential points of corrosion.

3.15.3 Rusting of rebars in concrete

As concrete ages, faulty design or improper curing results in inadequate drainage and in crack formation. Additional cracks result from the stresses of thermal expansion and contraction of the concrete. Seepage of water into these cracks, half a millimeter in size or larger, carries dissolved oxygen and carbon dioxide into the concrete.

This dissolved oxygen and the carbon dioxide are active in promoting corrosion. For example, oxygen maintains the alkalinity of the cathodic electrodes by its conversion to the hydroxide ion. This reaction keeps the pH at the cathodes at a high alkaline value. Dissolved carbon dioxide destroys the passivity of steel at the anodes by neutralizing the alkali in the hydrated cement and by releasing free chloride ions from calcium-chloride-containing cement (carbonation). Also, as water seeps through the cracks, it leaches out the soluble alkali from the concrete. As the pH of the concrete adjacent to the steel drops below 9, the protective film of ferrous oxide deteriorates and the passivity it once provided the steel surface is lost.

Pitting occurs at the anodes as iron dissolves to form ferrous ions. At the cathodes, oxygen and water are converted to hydroxide ions, which further increases the corrosion voltage. Thus the corrosion process is accelerated and serious structural deterioration occurs if further environmental exposure is not prevented.

3.15.4 Chloride-induced corrosion

The presence of chloride ions in concrete, above a critical value, causes steel to start corroding. The presence of these chlorides at the steel surface alters the anodic electrode potential. This prevents the formation of a protective coating of ferrous oxide. Since the steel surfaces are not passive, corrosion occurs at an accelerated rate.

Chlorides can originate in the concrete or they can enter from the environment. They occur in the concrete as free chloride and as bound chloride. Free chlorides in the freshly mixed concrete come from contamination of the sand and gravel aggregates and from the additives used. Bound chlorides are found in hydrated aluminum cement. Carbon dioxide is able to free these bound chlorides by converting this calcium chloride component of cement to calcium carbonate. Chlorides from the environment can enter concrete through cracks. Although the main source is from rain, snow, ice, fog, or ocean spray, chlorides can come from the use of deicing salt treatments of streets and highways.

3.15.5 Corrosion protection

There are three ways to limit the dangers of structural failure in steel-reinforced concrete.

The first method is to select noncorrosive materials suitable for the environment. The quality of the concrete is important, such as high alkalinity of the hydrated cement and an absence of chlorides in the mixed concrete. Other considerations are the thickness and curing of the concrete cover, proper location of the rebars to reduce any unnecessary stress-produced cracks, and good drainage. Flat surfaces that collect water are especially vulnerable. The use of corrosion inhibitors, such as sodium or calcium nitrites for cathode protection and sodium benzoate for anode protection, also creates problems. Galvanized and corrosion-resistant steel rebars have been tried, but usually are not cost-effective.

The second approach is to prevent corrosive materials from penetrating to the steel surface. The ideal solution is to cover the concrete with an impervious coating of asphalt, followed by a second coating of a more durable, traffic-resistant hard coating. This is difficult to achieve and is costly. In practice, cracks in the concrete surface are sealed as they appear. This requires constant maintenance. Numerous sealants are being developed for this purpose. One of the best appears to be an epoxy cement that is plasticized to match the expansion coefficient of the concrete.

The third solution is the use of anodic protection to prevent further corrosion of the reinforcing steel. There are two ways this can be done: either by using a sacrificial anode, or by applying a cathodic protective voltage with an electric generator. Both methods require the reinforcing steel to form a complete electric circuit. This means that the rebars must make electric contact with one another and the voltage source. Mapping techniques involve the measurement of voltage or resistance gradients to locate regions of serious anodic corrosion. This requires good instrumentation and qualified engineers.

3.16 Building Problems Related to Concrete

In general it is true that the most dense and impermeable concrete is more durable and more resistant to a hostile environment. Quality concrete requires carefully graded sizes of aggregate, adequate proportion of cement, selection of the appropriate type of cement for the intended application, and proper curing.

Some of the common building problems related to concrete are listed here with their causes and suggestions for correction.

1. Sulfate deterioration of concrete is caused by moisture and sulfate salts in the soil that is in contact with concrete foundations, floor slabs, and walls. Type V sulfate-resisting cement is made for this purpose. To correct the problem (if the soil cannot be kept away from the concrete), better drainage might keep the soil dry and the salts in solid, not solution, form.

2. Efflorescence is the appearance of an unsightly fluffy white crust on the surface of walls. It is caused by salts in solution (in the concrete, the stone, or the bricks) moving to the surface of an interior or exterior wall. As the water evaporates from the salt solution in dry weather, a loose mass of white, powdery salts remains. The process of efflorescence slowly continues as the soluble salts are leached from the concrete by heavy rain entering the wall through joints or cracks. These soluble salts can come from admixtures, but most likely will come from poorly washed aggregates. Some relief from efflorescence can be gained by treating the surface of the wall with a water repellent and sealing all cracks and joints to keep out rain. If no cracks or pin holes are in the coating, these soluble salts should no longer be able to reach the surface.

3. Freeze-thaw cracks (forming in concrete in subfreezing weather) can be caused by concrete with a water/cement ratio that is too large. This can produce tiny crevices and voids around the aggregates, allowing penetration of water into the concrete by wind-driven rain. Tremendous forces, produced by the expansion of water as it freezes to form ice, cause spalls and cracks in the concrete. Prevention requires a better water/cement ratio and an examination of the admixtures if any were used. Correction requires waterproofing the surface of the concrete with a polymer-modified cement-based surface coating. Further protection could be gained by applying a protective coat of paint.

4. Corrosion of steel rebars can cause cracks and rust stains to appear in the concrete. (Refer to Sec. 3.15.)

5. Leaks in concrete roofs or parking decks are due to water penetrating the surface. Although these surfaces should have been level, slightly raised in the middle, or slightly tilted toward the edge, rainwater often stands in pools. This water eventually penetrates into the concrete. Other cracks form from freezing and thawing or from heavy traffic on decks. Perhaps the parking deck is subjected to overloads

and should have had better reinforcement. Perhaps expansion joints should have been used. Water leaks eventually form in the ceiling, with efflorescent salts showing in the cracks. A waterproof coating alone rarely works because the cracks continue to grow. A solution is to use epoxy in the clean cracks and to fill them with a flexible sealant material.

The economic pressure on the building industry to reduce construction costs often results in high maintenance costs and expensive repairs. Designs often leave too little margin for overloading and other abnormal use of structures. In addition to unusual environmental conditions, natural disasters such as floods, mud slides, hurricanes, tornadoes, record-breaking hot and cold weather, and earthquakes are a constant threat. In the meantime we find that insurance is becoming increasingly difficult to obtain and expensive. All these problems, when combined with tight construction schedules, often result in substitutions and short-cut innovations which invite early deterioration and failure. In addition, there is pressure to use faster construction techniques and materials. The satisfactory performance of materials used in building construction poses a challenge to workers who must become more highly trained and maintain a close interaction between design and construction. Recognizing the merits of the professional performance of trained workers is long overdue.

Ultimately, the consumer is challenged to become more knowledgeable of the building profession and construction practices. Where practical, the owner should be involved in the care and maintenance of the property. World ecology urges an end to "planned obsolescence" and misuse of our natural resources.

3.17 Health Hazards

For some time cement plants, busily grinding clinkers to make cement, have been enclosing the grinding mill to keep dust from getting into the outside air. Unslaked lime evolves heat when mixed with water and is a caustic irritant to the skin or lungs. Wet cement paste can dry the skin and cause alkali burns. Dry cement dust can cause inflammation of the tissues of the eyes and nose. Prolonged exposure to cement dust can cause dermatitis and delayed silicosis. Workers must use protective respiratory devices, eye protection, and gloves. OSHA regulations must be followed carefully.

Terms Introduced

Accelerating agents
Admixtures
Asbestos fibers
C_3S
Cement symbols
Cementitious material
Clinker
Concrete admixtures
Concrete aggregates
Concrete air-entrainment agents
Concrete alkalinity
Concrete composites
Concrete durability
Concrete porosity
Concrete sulfating
Efflorescence
Gypsum plaster

Hydrophilic
Lime setting
Passive steel
Plaster of paris
Polymer concrete
Portland cement hydration
Prestressed concrete
Quicklime
Rebar corrosion
Rebar passivity
Reinforced concrete
Salt corrosion
Slaked (hydrated) lime
Slump test
Water-reducing agents
Water/cement ratio
Whitewash

Questions and Problems

3.1. Why are reactions with solids slower than those between ions in solution?

3.2. What are the distinguishing features of the five types of cement that determine their selection?

3.3. What are the chief ingredients of concrete and the purpose of each?

3.4. How do lime and plaster of paris differ in their setting process?

3.5. Why does cement take so long to get its final strength?

3.6. Briefly describe the stages that cement goes through in setting and hardening.

3.7. What are the common types of admixtures and what do they accomplish?

3.8. Why is calcium chloride restricted in use as an accelerator in concrete that contains reinforcing steel?

3.9. What causes efflorescence?
3.10. What types of composites are used in concrete?

Suggestions for Further Reading

James T. Dikeou and David W. Fowler, eds., *Polymer Concrete: Uses, Materials, and Properties,* American Concrete Institute, Detroit, Mich., 1985.

David W. Fowler and Lawrence E. Kukacka, *Applications of Polymer Concrete,* Publ. SP-69, American Concrete Institute, Detroit, Mich., 1981.

Peter Mendis and Charles McClaskey, *Polymers in Concrete: Advances and Applications,* Publ. SP-116, American Concrete Institute, Detroit, Mich., 1989.

V. S. Ramachandran, *Concrete Admixtures Handbook (Properties, Science, and Technology),* Noyes, Park Ridge, N.J., 1984.

W. H. Ransom, *Building Failures: Diagnosis and Avoidance,* E. & F. Spon, London, 1987.

M. R. Rixom, *Chemical Admixtures for Concrete,* 2d ed., E. & F. Spon, London, 1978.

P. Schiessl, *Corrosion of Steel in Concrete,* Chapman and Hall, London, 1988.

J. F. Shackelford, *Introduction to Materials Science for Engineers,* Macmillan, New York, 1985.

R. N. Swamy and B. Barr, *Fibre Reinforced Cements and Concretes: Recent Developments,* Elsevier, London, 1989.

Joseph J. Waddell and Joseph A. Dobrowolski, *Concrete Construction Handbook,* 3d ed., McGraw-Hill, New York, 1993.

Don Watson, *Construction Materials and Processes,* 3d ed., McGraw-Hill, New York, 1986.

Chapter

4

Masonry

4.1 Introduction	92
4.2 Early Masonry	92
4.3 Modern Masonry	93
4.4 Masonry Mortar	95
4.5 Ingredients of Mortar	97
4.5.1 Portland Cement Used in Mortar	97
4.5.2 Blended Cement	97
4.5.3 Masonry Cements	98
4.5.4 Lime	98
4.5.5 Aggregates	98
4.5.6 Water	99
4.5.7 Admixtures	99
4.6 Physical Properties of Mortar	99
4.7 Thickness for Optimum Bond Strength	100
4.8 Grout Masonry	100
4.9 Concrete Structures	101
4.10 Masonry Units	101
4.11 Concrete Blocks	102
4.11.1 Classification	102
4.11.2 Manufacture	103
4.11.3 Aggregates	103
4.11.4 Admixtures	103
4.11.5 Properties	104
4.12 Cement Bricks	104
4.13 Clay Masonry Units	104
4.13.1 Classification of Clay Units	104
4.13.2 Clay Bricks	105
4.13.3 Clay Tile	106
4.13.4 Terra Cotta Tile	107
4.14 Stone Masonry	107
4.15 Glass Blocks	111
4.16 Plaster	112
4.17 Restoring Masonry Surfaces	112
4.17.1 Water Cleaning	113
4.17.2 Abrasive Cleaning	115
4.17.3 Chemical Cleaning Agents	115

4.18 Hazardous Chemicals	116
4.19 OSHA Responsibilities for Workers	117
4.20 Masonry Problems and Their Prevention	117
4.21 Environmental Hazards	118
Terms Introduced	118
Questions and Problems	118
Suggestions for Further Reading	119

4.1 Introduction

Stone structures are some of the oldest and most durable structures of our heritage. In this chapter we trace the evolution of masonry from the early stone cathedrals to the modern technology of stone, clay, concrete, and glass masonry units. These and other related topics described in the chapter include steel-reinforced block structures, grouting, plastering, and restoring surfaces of masonry walls.

4.2 Early Masonry

Historic cathedrals, stone bridges, and walls testify to the early development of stone masonry (Figs. 4.1 and 4.2). The Romans are credited with perfecting the design of the large cathedral arches and domes. They distributed the lateral thrust of the load of fitted stones to the walls and their buttresses. Many retaining walls built by the Romans in Britain during the first century are still standing and useful.

Wattle and daub walls of homes in Britain have been standing for centuries. Walls are constructed of interwoven willow wands filled with dung and mud and packed between timbers. Homes undergoing restoration must retain these valuable centuries-old walls. By order of Queen Victoria, the wattle and daub is painted white and the timbers are dark brown or black (Fig. 4.3).

Brick, also developed and used centuries ago, is one of our oldest human-made building materials.

With the discovery of portland cement in the nineteenth century, the construction of multistory commercial buildings became possible. The improvements in steel-reinforcing bars further increased the use of reinforced concrete. Consequently, the use of stone masonry became limited to the decorative facings on buildings. Attention was drawn away from stone masonry, thereby retarding its development.

Figure 4.1 Wall built by the Romans in the first century for the defense of Chester, England.

4.3 Modern Masonry

During the mid-twentieth century, masonry came back into favor. In addition to adopting the advances made in concrete construction, the use of masonry spread to other building materials. Masonry units may be defined as any type of small, solid, or hollow units of building material that are held together with mortar. These units usually include stone, cast stone, cement brick and concrete block, clay brick and tile, and glass blocks.

Vital to the successful performance of each of these masonry unit systems is the selection of the proper mortar to hold the units together and keep out the weather. Since no machine has yet been invented to assemble the masonry units in place, the performance of the masonry structure depends on the quality of the mortar, the skill of the mason, and exposure to the environment. Severe environmental conditions, such as torrential rains and intense sunshine, require more careful design and higher-quality ingredients than do the more protected environments. This is especially true if these masonry structures are to withstand earthquakes or hurricanes.

Figure 4.2 Built during the thirteenth century, Salisbury Cathedral, Salisbury, England, with one of the tallest steeples in Europe, is an outstanding example of early masonry.

Figure 4.3 Dating from the early fifteenth century, this two-story house retains the wattle and daub construction on the front wall. The roof is covered with stone tile native to the area. Mason's Court, Stratford-on-Avon, England. (*Courtesy of Ann Brandman.*)

4.4 Masonry Mortar

Less than 1 percent of the weight of masonry structures consists of the mortar holding them together. Cement, hydrated lime, aggregates, and water are the necessary ingredients that make this feat possible.

Table 4.1 lists the composition (by volume) and the minimum compressive strength for the common types of masonry mortar. Ranges in

TABLE 4.1 Composition and Strength of Mortar

Mortar type	Cement*	Hydrated lime*	Aggregate*	Compressive strength,† psi
M	1	¼	2¾–3¾	2500
S	1	¼–½	2¾–4½	1800
N	1	½–1¼	3¼–6¾	750
O	1	1¼–2½	5–10½	350
K	1	2½–8		75

*Parts by volume of portland or masonry cement.
†After 28 days.

composition are given to allow the choice of mortar to fit more closely the conditions of the structure and, perhaps, reduce some of the costs, provided, of course, that satisfactory strength and performance are obtained. (The ASTM standard for mortar for unit masonry is C 270.)

The five basic types of mortar are known as types M, S, N, O, and K. In contrast to concrete, each type contains some hydrated lime. For economy, it is a general rule that mortars need have no more compressive strength than the masonry units they hold together since the masonry system is no stronger than its weakest link.

Type M mortar is a high-strength all-purpose mortar with good durability. Generally recommended for exterior walls and reinforced masonry where high strength is required, it is designed for below-grade use in foundations, sidewalks, and retaining walls.

Type S mortar has slightly less strength and durability than type M. Designed for masonry systems requiring strong bond strength and resistance to lateral movement, it is suitable for reinforced hollow-block masonry.

Type N mortar is a medium-strength general-use mortar recommended for use above grade and able to withstand the environmental exposure required of exterior and parapet walls which receive severe exposure.

Type O mortar, a low-strength product designed for interior use, is suitable for non-load-bearing masonry where freeze-thaw temperatures do not occur.

Type K mortar is seldom mentioned in specifications for mortar. For example, it is not listed in ASTM C 270. Type K is a very low-strength mortar with use restricted to non-load-bearing interior walls and minor exterior use where minimum height and minimum weather exposure are expected. Economy is its chief attraction.

The strength of mortar is shown in Table 4.1, indicating an increase in the proportional amount of cement. With such a wide variety of mortars and overlapping uses, it should be clear that no all-purpose mortar is available. The choice of the mortar requires a careful assessment of the kind of use and the environmental exposures the masonry system can expect. Even then the choice of mortar and masonry unit will need to be a tradeoff between the special properties of each mortar, its cost, availability of materials, and the number of skilled workers in the region. For example, many building codes allow use of either type M or S mortar in masonry walls. Type S mortar is often used because of its superior workability and other properties.

4.5 Ingredients of Mortar

Four ingredients are essential to the satisfactory performance of mortar. Cement provides mortar with the necessary strength; hydrated lime provides the elasticity and water retention so necessary for workability; sand provides durability and strength in addition to acting as a filler; and an optimum amount of water is necessary for good bonding, plasticity, and workability. Selecting the correct ingredients is important for the optimum performance of the mortar.

4.5.1 Portland cement used in mortar

Portland cement occurs as finely ground powders available in five types (see Table 3.2). (Detailed specifications for portland cement are found in ASTM C 2150.) These five types of portland cement differ from one another primarily in the proportions of the four compounds: dicalcium silicate, tricalcium silicate, tricalcium aluminate, and tetracalcium aluminoferrite. When water is added, it reacts chemically with the surface of the cement particle to slowly form gelatinous hydroxides, which gradually cement the sand particles together.

The choice of cement type to use in the mortar depends on the circumstances, but use is usually limited to type I, II, or III.

4.5.2 Blended cement

Blended (hydraulic) cement is listed in ASTM C 270 as an alternative to portland cement. Fly ash and slag are examples of cementitious materials which qualify for use in a blended cement. These waste materials are by-products of steel smelters. They become cementitious when mixed with water because of the large amount of calcium hydroxide produced in the reaction.

With an increasing need to conserve natural resources, the use in concrete of fly ash and slag has become urgent. Their use in blended cement mixtures saves the energy used in manufacturing portland cement as well as the substitution of a waste material for portland cement. The specifications for blended hydraulic cements are given in ASTM C 595 for slag or pozzolan with portland cement.

Pozzolan cement is burnt clay or shale and resembles the volcanic dust found in Italy. Pozzolanic materials are not cementitious since they contain less calcium than fly ash and slag. However, in practice this is remedied by the addition of calcium hydroxide. The chemical reaction of water and added calcium hydroxide converts the silicates and the aluminates in them into cementitious cements.

4.5.3 Masonry cements

Masonry cement is a proprietary material. It is given as an option in place of portland or blended cement for M, N, or S mortar in ASTM C 270.

The choice of using portland cement or a premixed masonry cement is usually one of convenience and cost. The composition of masonry cement varies with the manufacturer, but it may contain such admixtures as calcium salts, set retardants, or air-entraining agents. (ASTM C 91 gives the specifications for masonry cement.) Specifications usually do not limit the admixtures except to warn that the use of calcium chloride may have a detrimental effect on metal. This salt has a very strong corrosive effect on ferrous metals such as steel rebars.

To obtain the desired plasticity needed in mortar, masonry cements may contain blends of portland cement, masonry cement, and hydrated lime. The mixtures may also contain some slag, hydraulic lime (calcined limestone), or powdered limestone. Proprietary masonry cement mixtures should be tested individually for the desired properties.

4.5.4 Lime

Lime is a general term referring to calcium oxide, calcium hydroxide, or even hydraulic lime. These forms of lime have distinctly different properties. (The specifications for lime are given in ASTM C 207.) Calcium oxide (CaO) is known as quicklime or unslaked lime. Calcium hydroxide [$Ca(OH)_2$] is known as slaked lime or hydrated lime. Hydraulic lime, an impure form of calcium oxide, is obtained by burning hydraulic limestone. Calcium oxide chemically combines with water to form calcium hydroxide with the evolution of considerable heat. Each one of these limes is irritating to the skin, but calcium oxide is classified as a strong irritant. Therefore calcium hydroxide should be used, whenever possible, to avoid the danger of using calcium oxide. Hydrated lime provides the needed plasticity to improve workability of mortar.

4.5.5 Aggregates

The size of sand particles, and the relative amount of each size used, is critical to the performance of mortar. Since a relatively thin layer of mortar is used to cement the masonry units into place, a careful assortment of sand size is necessary to provide workability and a good bond. (ASTM C 144 lists the specifications for aggregates for mortar.) There must not only be a graded range of particle sizes, but the size must fall between very fine sand and sand of 0.6-centimeter (¼-inch) diameter. A load of sand deficient in the amount of fine sand would lack workability. A lack of the larger sand particles lowers its durability and strength. Aggregates in mortar should be closely packed with the smaller particles fitting into the voids between larger particles in the matrix of the cement paste.

4.5.6 Water

Mortar needs enough water to form a paste that can be easily spread with a trowel, but does not slump. Yet the mortar must have enough water to be fluid enough to displace air from cracks and crevices between the masonry units. To provide good workability and bonding, the maximum amount of water allowed in the mixture should be used. As with concrete, the water/cement ratio is very important for strength as well as workability.

The size of the batches of mortar should not exceed what can be used in 2 or 3 hours. However, if mortar begins to dry out and stiffen in less than 2 or 3 hours, it is permissible to carefully mix in additional water to restore workability.

4.5.7 Admixtures

Until recently, hydrated lime was the only admixture used in mortar. Now a vinyl polymer is used to provide workability and improved bonding, particularly for patching and repairs. A wetting agent which can act as an air-entraining agent is sometimes used, especially in proprietary mortar mixes. Sodium laurylsulfate, a common kitchen detergent, is added to mortar as an air-entraining agent. This lowers the surface tension of water, allowing tiny air bubbles to be more lasting. These minute bubbles, distributed throughout the mortar, serve as a lubricant and make the mortar more workable. Nevertheless, excess air bubbles are to be avoided as they reduce the bond strength.

4.6 Physical Properties of Mortar

In some ways the properties of mortar are more critical than those for concrete. Compressive strength is one of the main assets of cured concrete. In addition to compressive strength, mortar must have adequate bond strength, shear strength, and durability. Successful performance depends on its workability and its skillful application.

Workability, one of the most essential properties of mortar, determines the success of its application. This is provided by slaked lime, careful gradation of aggregates, and the proper amount of water. Slaked lime provides water retention, elasticity, and workability, and in its absence the mortar is stiff and difficult to use. Because of the subjective nature of workability, it is challenging to define but is easily recognized by a skilled mason.

Mortar must have a strong bonding strength, which requires that the mortar be able to flow into crevices and small voids. For the mortar to set properly, it needs water retention to avoid too much water being sucked out of the mortar gel by the porosity of the masonry unit. These essential properties are usually provided by the slaked lime.

Such inorganic materials as mortar can contain several forms of water, such as water of hydration, water contained in a gel-like paste, and free water. Water of hydration is part of the compound. To release the water, the compound must usually be decomposed by heating. Gypsum ($CaSO_4 \cdot 2H_2O$) is an example of such a compound. Partially dehydrated gypsum (plaster of paris) is often used in patching grout and mortar. An example of bound water is the gel-like structure formed by the hydration of cement. An example of excess (free) water is the water that can be squeezed from mortar to "float" the masonry unit. This excess water is to be avoided as it reduces the bond strength holding the masonry unit in place. Mortar should have just enough water to form a workable paste.

Water retention, provided by using slaked lime, is necessary to avoid the problem of some concrete masonry units, namely, sucking water from the mortar. Also, on a dry and hot summer day, water can evaporate from the mortar before it forms a good bond. This loss of water can reduce the workability of the mortar seriously by reducing its flow and plasticity.

4.7 Thickness for Optimum Bond Strength

A general rule for an adhesive is: the thinner the layer of the bonding mixture, the stronger the bond. This is in sharp contrast to the popular belief that if a little is good, more is better. It is best to use only enough mortar to fill the irregularities in the surface of the materials being held together. Most adhesive failures are due to flaws or imperfections in the adhesive layer itself.

4.8 Grout Masonry

After a masonry structure is completed, grout is used to fill in the remaining crevices and joints. Grout differs from masonry mortar in its fluidity since it is poured and not spread into place with a trowel. Grout also needs finer aggregates. (Specifications for grout for reinforced and nonreinforced masonry are given in ASTM C 476.) Masonry grout is essentially composed of portland or blended cement, fine or coarse sand, water, and a small amount (if any) of calcium hydroxide. Grout is identified as fine or coarse grout, depending on the size of the sand aggregate used.

Grout is used in reinforced load-bearing masonry walls. For example, it is used in large dams to fill the cracks formed as the concrete cools. In other hard-to-reach places, such as the spaces between tunnel walls and the surrounding earth, grout is used to spread the earth stresses uniformly over the structures. Grout is also poured, or pumped, into building structures—for example, around the steel-reinforcing rods in hollow concrete blocks.

4.9 Concrete Structures

Structural masonry is divided into load-bearing, non-load-bearing, and decorative veneers used on walls of buildings. Concrete masonry includes the assembly of walls of solid or hollow units. They may be reinforced or nonreinforced and interior or exterior walls. The walls may contain clay, tile, or glass units.

Load-bearing concrete structures are usually steel-reinforced.

Non-load-bearing concrete structures include solid and hollow walls, decorative veneer walls, walks, patios, retaining walls, and interior wall partitions.

Concrete walls can provide thermal and sound insulation and fire protection.

4.10 Masonry Units

Masonry units are composed of stone, cement, clay, or glass and are made in hollow or solid blocks. Clay bricks can be either load-bearing or non-load-bearing.

Masonry units are used in solid walls, cavity walls, and steel-reinforced walls. Selection of the grout to use depends on environmental exposure, building type, building height, and loading requirements.

For load-bearing walls the masonry structure requires a careful design by engineers or architects to take full advantage of the technological advances made in steel-reinforced concrete buildings. Masons use many of the latest technological advances developed for reinforced concrete buildings.

Solid nonreinforced walls are composed of solid or hollow masonry units and support only a light load. Solid structural walls are held together with joint reinforcement, metal ties, or masonry headers. Masonry-bonded walls use headers or have the blocks overlapping in alternate courses.

The type of wall selected depends on the severity of the exposure to environmental factors, such as the possibility of strong winds and penetrating rain. In severe weather conditions, including wind-driven rain, a coating of stucco, sealed with paint, is recommended for waterproofing the concrete-block walls.

Reinforced walls using hollow concrete blocks can be one or two wythe (masonry units) thick. If one wythe (single block width) is used, the steel-reinforcing rods are placed into the core of the hollow block, which is then filled with grout. In double-wythe construction the rebars are placed between the two wythes and grouted into place.

4.11 Concrete Blocks

Concrete blocks can be solid or hollow, but the hollow 8- by 8- by 16-inch (20- by 20- by 40-centimeter) blocks are the most common. They can be load-bearing or non-load-bearing. (The ASTM standard specification for hollow load-bearing concrete blocks is ASTM C 90.) The three types of blocks are classified as being of normal weight, medium weight, and lightweight, depending on the weight of the aggregate contained.

4.11.1 Classification

Type I concrete blocks have been stored in an environment with controlled humidity to be in moisture equilibrium with the environment in which they are to be used. Concrete blocks not matched to the humidity of the building site are known as type II. Grade N is used to designate the all-purpose concrete blocks. This grade is suitable for exterior walls and for use above or below ground, and acts as a barrier to moisture penetration. Grade S concrete blocks are restricted largely for use in constructing interior walls.

Thus grade N-I concrete blocks are all-purpose blocks that have been seasoned in the humidity which matches their region of intended use. Grade S-I concrete blocks, intended for interior use, have also been humidity-matched to the region in which they will be used. By contrast, grade N-I concrete blocks have been humidity-aged for the region of use and are limited to use above ground.

4.11.2 Manufacture

Most concrete blocks and cement bricks are cured in steam kilns at atmospheric pressure. In this process the time for the concrete to reach full compressive strength is reduced from the usual 28 days to 24 hours. Hollow blocks are machine-formed at a rate of about 1000 units per hour. They are composed of portland cement, graded mineral aggregate, and just enough water for the mixture to hold its shape. There are two methods of curing these blocks. One method uses a kiln, heated with steam at atmospheric pressure to about 65°C (150°F) for 18 hours. The other method uses a kiln heated with pressurized steam to about 177°C (350°F) for 12 hours.

The cured blocks are then stored under controlled humidity until their moisture content reaches that of the environment in which they will be used. This is necessary to avoid the blocks pulling apart as they continue to dry after being set into place at the building site. The combined shrinkage of the blocks, as they lose moisture, can cause cracks to form and seriously weaken the structure. This can be a problem in dry climates where the relative humidity is below 50 percent.

4.11.3 Aggregates

Aggregates usually consist of sand, burned clay, cinders, slag, and stone. Although fly ash, slag, and powdered silica are classified as aggregates, they slowly react with the lime produced by the hydration of cement to provide additional cementitious materials. This further increases the strength of the concrete as it cures. The aggregate-size distribution improves concrete strength and contributes much to the performance of the concrete units just as the size distribution of aggregates in mortar is important.

4.11.4 Admixtures

The use of admixtures provides concrete blocks with the properties needed for the particular regional environment. For example, surfactants, which lower the surface tension of the water in the concrete, act as air-entraining agents. They are used to promote the formation of tiny stable air bubbles, which help the blocks to withstand the freeze-thaw cycles in cold climates. These cycles reduce the strength of the concrete blocks.

Another rather ingenious way to form tiny bubbles is to mix aluminum powder with slaked lime before the mixture is formed into concrete blocks. The aluminum powder chemically reacts with the lime to form hydrogen gas and calcium aluminate. The pressure of the hydrogen gas produced in the heated mold causes the concrete mix to completely fill the mold. No reference is made in the literature about the explosive nature of hydrogen gas with oxygen from the air.

4.11.5 Properties

The properties of concrete blocks depend on the type of cement and the size of the aggregate used. The larger the proportion of cement used, the stronger the concrete. Curing the concrete with steam produces in 1 day the strength normally obtained in 28 days at ambient temperatures. This is to be expected because the rates of these exothermic hydration reactions generally increase as the temperature is increased. In addition to the rapid production of concrete blocks, there is greater strength and less danger of cracking. This is due to the carefully controlled humidity during setting and curing, which reduces the tendency of the concrete to shrink. Shrinking usually occurs from fluctuations in moisture content experienced during a 28-day curing period. Although proprietary admixture products are available, it is doubtful whether they are really needed or used.

4.12 Cement Bricks

Cement bricks are made from portland cement, selected aggregates, and water by a process similar to that used for concrete blocks. There are two types of cement bricks. Like concrete blocks, type I is moisture-controlled and type II is not moisture-controlled. There are also two grades: grade N is an all-purpose quality, whereas grade S is less resistant to moisture penetration, resulting in less of a freeze-thaw challenge. Both types I and II are available in grades N and S. (Specifications for cement building bricks are given in ASTM C 55.)

4.13 Clay Masonry Units

Some of the most durable building materials are made of clay: bricks, ceramic tile, and terra cotta. In contrast to the concrete masonry units, which depend on the ingredients reacting with water, clay masonry units are heated until the clay melts and flows over the surface of the aggregates. This bonds the aggregates together and forms an impervious, vitreous ceramic material with good compressive strength.

4.13.1 Classification of clay units

Although both facing brick and building brick have high compressive strength, building brick (common brick) is separated by ASTM into grades depending on the severity of the exposure it is to withstand.

Grade SW clay brick is made for use below or above ground level and in regions where severe weathering conditions are found. It must resist the penetration of moisture where freeze-thaw conditions occur.

Fire clay, using the stiff-mud process and baked in a heated kiln, produces clay masonry units with high compressive strength and low moisture absorption. Since it is the moisture in the clay that expands when frozen, low moisture absorption provides the baked clay with greater durability and higher resistance to freeze-thaw cycles.

4.13.2 Clay bricks

The type of clay selected and its processing determine the structure and characteristics of clay units in building structures (Fig. 4.4). Most bricks and structural tiles are made by the stiff-mud process with 12 to 15 percent moisture content providing the needed plasticity.

Natural clay, and especially the mineral bentonite, has the thixotropic property of becoming more fluid when stirred, but stiffens when the stirring stops. This structural property, when stirred, is due to the alignment of its tiny platelets. On standing, they quickly become disordered due to thermal agitation. Later, after deaerating in a partial vacuum to remove excess bubbles, the formed clay is shaped and cured by a burning process.

After partial drying, the molded clay tile or brick unit is kept under controlled temperature and humidity for a day or two. The clay units are slowly burned at 982 to 1315°C (1800 to 2400°F) for 1 or 2 days. The loss of moisture is a slow process in a gel and cannot be hurried by raising the temperature. Increased temperature would seal off the surface and force the water to gush out of the clay to form a porous solid rather than giving the water time to diffuse slowly through the gel to the surface.

Clay bricks must be laid in place with care to obtain a secure bond with the mortar. These bricks, unlike concrete masonry units, are not delivered at the job site conditioned to the humidity of the surroundings. They absorb water from the mortar by capillary attraction and, thus, dehydrate the mortar. To avoid this problem, the bricks are soaked with water and left to dry to the ambient humidity conditions. If the surface is still wet, the bricks will float and fail to form an acceptable bond. Over time clay bricks will expand, while under the same conditions, concrete will shrink. Both of these stress conditions

Figure 4.4 View of interior brick walls and arches in the city hall, Stockholm, Sweden. Stairway leads to the banquet room where Nobel recipients are honored.

can result in cracks or poor bonding. The solution is to provide expansion joints for brick structures.

4.13.3 Clay tile

Clay tiles are often used as facing that is anchored to the structural steel framing of the building. Although its popularity has diminished, the use of clay tile for restoration purposes continues. In addition to its use for load-bearing and non-load-bearing wall structures, it is used for floors and interior walls.

Tiles are made in the same manner as clay bricks: by firing them in a kiln to a temperature high enough to melt the silicates in the clay so they can cement the aggregate particles together in a ceramic matrix. Clay facing and glazing tile has an attractive finish that can serve as an interior wall without the expense of painting the surface.

There are two grades of structural clay tile for exterior exposure. The ASTM designation LBX is reserved for the highest grade, suitable for the most severe exposure to moisture and freezing temperatures; for less severe exposures the LB grade may be used. ASTM C 34 handles both grades of clay tile.

Ceramic glazed facing tile is a load-bearing tile unit that has a finish impervious to moisture. The units are available in grade S and a select grade SS with ground edges and closer dimensional tolerances. For type I only one face of the tile is glazed. The tiles are classified as type II if both sides are to be exposed and need to be glazed. Standards for glazed brick as well as glazed tile are described in ASTM C 126.

4.13.4 Terra cotta tile

Although terra cotta tile has been used for centuries, dating back to the days of the Romans, it is no longer popular and is used mainly in restorations. The term "terra cotta" stands for fired earth. This tile is included in the specifications for clay brick and masonry.

4.14 Stone Masonry

The stones that qualify as building materials can be classified as granite, limestone, coral, sandstone, slate, marble, and lava. The use of stone and slate is illustrated in Fig. 4.5. These can vary greatly in compressive strength—between varieties of stones and between stones from the same source. This is due to the complexity of their compositions and wide variations in the percentage of mineral components. Even though stones are abundant throughout the world, most of them are unsuitable for building purposes. They must be accessible, easily quarried, and satisfy the stringent requirements of appearance, durability, hardness, porosity, and workability.

Granite, used as a building material since the beginning of civilization, is a visibly crystalline igneous rock with granular texture and composed of quartz, feldspar, mica, and hornblende. It is used extensively in the interior and exterior of government and other office buildings. Depending on the proportion of the minerals, the colors vary. They may be pink, red, brown, buff, cream, gray, dark green, or black.

Limestone, a sedimentary rock composed mainly of calcium carbonate or magnesium carbonate, is durable, workable, and distributed throughout the earth's crust. Fossilized remains of animals (fish, shells, coral) and plants are evident in most limestone. Colors may be white, cream, buff, gray, or variegated patterns of all of these colors. Limestone, easily worked when first quarried, later develops a hard, long-wearing surface after exposure to air. Travertine, a form of limestone found in deposits at the mouth of a hot spring, can be polished and often resembles marble.

Figure 4.5 This historic home is constructed of stone and slate common to the Lake District, Grasmere, England.

Coral limestone, composed of reef-forming coral often of great extent, consists chiefly of calcareous skeletons of corals, coral sands, and the solid limestone resulting from their compaction. This coral limestone forms on the ocean floor bordering the shores of islands and lagoons. Plentiful, inexpensive, and used often in the eighteenth and nineteenth centuries in the Hawaiian Islands, coral is now quarried very carefully due to ecological concerns and has become expensive. Now it is most often used as veneer. Coral reefs can be renewed, slowly, and are living things to be protected. The beautiful Kawaiahao Church, a national historic landmark in Honolulu, Hawaii, was built in 1839 of blocks of white coral (Fig. 4.6). Today, though the damage is hardly noticeable, those thick walls of coral are being pecked by birds looking for salt.

Sandstone, a sedimentary rock composed of individual sand or quartz grains held together by cementitious material, contains a high degree of iron oxide which gives it a red or brown color. It is soft and easily worked when first quarried due to a high water content. After the water evaporates, it hardens but is not always durable. Varying in color from buff, cream, pink, red, blue-gray, to brown, sandstone lends itself to finishes such as veneers, sills, coping, and moldings (Fig. 4.7).

Slate, a group name for various fine-grained rocks derived from mudstone, siltstone, and high-silica clays and shale sedimentary deposits, is characterized by planes which easily split into thin sheets and lines. Its use as school blackboards is being replaced by new plastic-surfaced boards (chalkboards) of various colors. Slate roofing tiles are used extensively in Europe.

Marble, a metamorphic crystalline limestone composed of calcite or dolomite, is highly polished for commercial uses. Marble has been used for structural purposes throughout the centuries. In modern practice, marble is used as a beautiful interior and exterior veneer over a structural framework. It is used extensively for walls and columns, flooring, partitions, and decorative surfaces. American marble ranges from white to black; imported marble comes in hundreds of colors and patterns.

Lava is a crystalline or glassy igneous rock formed by the cooling of molten rock from volcanic vents and fissures. This rock can be very dense and heavy, or light in weight and bubbly. Colors range from light gray to dark red and browns, depending on the iron content. It is highly desirable and often used where lava is available. Seen extensively in Hawaii, the beautiful dark red, now costly "moss rock" walls are usually constructed without the mortar being evident—in retain-

Figure 4.6 Built of coral in 1839, Kawaiahao Church, Honolulu, Hawaii, has changed little over the years.

Figure 4.7 Modern decorative stone veneer.

ing walls, free-standing walls, as veneer over concrete block or wood, and in carvings. The older sections of Honolulu and the University of Hawaii-Manoa campus have beautiful dark brown lava curbstones that were put into place years ago when lava was quarried extensively and labor was inexpensive.

4.15 Glass Blocks

Produced in a variety of sizes, colors, and surface textures, glass blocks may be solid or hollow. In addition to their aesthetic appearance, their high transition of light somewhat reduces the need for interior illumination during daylight hours. The strength of glass blocks is much lower than that of most masonry units.

Glass blocks are not intended to be used as load-bearing units. They are used in a single-wythe (one-block-thick) wall and are usually laid with type N or S portland cement mortar. The permitted height and reinforcement depend on local municipal codes. Hollow glass blocks can be used as thermal and sound barriers.

4.16 Plaster

Plaster, applied by hand or by machinery, refers to the finished cementitious coating used on the exterior and interior walls of buildings to provide a smooth, finished appearance. This section is concerned with portland cement plaster, although lime or gypsum are sometimes used as the cementitious base of a plaster for wood structures.

Composed of cement and a plaster-grade aggregate—and, as an option, slaked lime—plaster should have a consistency appropriate to its method of application, have good durability, and withstand most kinds of weather. As in mortar, the slaked lime provides plasticity and the needed workability. To avoid cracks, the plaster must contain an aggregate of sand with a particle size less than 0.32 centimeters ($\frac{1}{8}$ inch) in diameter. The gradation in size of the sand is also important.

An external cement plaster, called stucco, is much used in mild climates. Plaster is as strong and durable as concrete and can be considered a modified form of concrete mortar. If used to cover wood, a metal screen or lath is attached to the wood surface to hold the plaster coating in place.

To avoid the need for papering or painting, the plaster may contain a mineral pigment or have a textured surface.

4.17 Restoring Masonry Surfaces

The cleaning and the restoration of exterior surfaces of buildings are important for proper appearance, maintenance, repair, and preservation. Extensive damage is caused by chemicals in the dust settling onto the surfaces of our buildings and by corrosive gases polluting the air of our cities. Although stone and concrete are strong and durable, surface porosity makes them vulnerable to the penetration of surface salts and acids accumulating on the buildings from the surrounding air. The action of the salts gradually causes the building surfaces to crack and crumble.

Much like the destructive freeze-dry cycles that occur in winter, the summer environment creates its own problems with expansion-contraction and wet-dry cycles. Water from light rains or the nighttime dew dissolves the salts in the dust that covers the buildings' exteriors. The porosity of the stone or concrete sucks the solution deep into the surface. During the heat of day, the salts crystallize to form a solution; these salt crystals occupy a larger volume than the water solution. The resulting stresses in the surface cause tiny cracks to form. These cracks grow in size and slowly crumble the surface as the wet-dry cycle is repeated each day.

As the salts are drawn deeper into the surface, some of them separate from solution as hydrates. Some, such as $Na_2SO_4 \cdot 10H_2O$, have a transition temperature of 32.4°C (90.3°F). This is the temperature at which the hydrate spontaneously decomposes into the anhydrous crystalline form of sodium sulfate (Na_2SO_4). Wide daily or seasonal fluctuation in temperatures would change these salts from their hydrated form to the anhydrous form. Again, the cyclic expansion and contraction creates stresses that cause cracks to develop, thereby weakening the internal structure of the masonry material. Therefore the salts act in two ways to weaken the stone or concrete building material.

After many years of dirt has become encrusted on a building surface, cleaning and restoring is not a simple matter. Many English churches and buildings (Westminster Abbey, for example) have been subjected to major cleaning (Fig. 4.8). Some cleaning chemicals can etch the surface as they clean, making the surface more porous and more difficult to clean in the future. Actually, the centuries of soot and other accumulations deposited on the surface serve as a protective coating, even though considered, by some, to be unsightly.

Three general cleaning methods to be considered are water cleaning, abrasive cleaning, and chemical cleaning, each with its advantages and disadvantages. Before cleaning can begin, preliminary investigations are needed to determine the types of masonry surfaces to be cleaned, the nature and extent of soil to be removed, and the most gentle method of removal that can be used efficiently. Many varieties of proprietary cleaning agents are on the market.

Masonry cleaning is a specialized service requiring skilled workers. It is usually necessary to seek out a reputable contractor to avoid seriously damaging the surface of the structure.

4.17.1 Water cleaning

Cleaning a building with water is often more forceful and prolonged than the normal exposure to wet weather and environmental conditions. To prevent interior water damage due to water leaks through the walls, it is necessary, first, to treat all visible leaks and joints. The process should be tested on selected areas of the building.

The high-pressure water-jet method (for cleaning and rinsing) uses adjustable pressures of up to about 1.389×10^7 pascals (2000 pounds per square inch) of pressure. Chemical cleaning agents can be introduced into the cleaning water. Lower pressure of up to 0.69×10^7 pascals (1000 pounds per square inch) is used if the surface is fragile. Higher pressures can strip the surface down to a fresh underlying

Figure 4.8 The recently cleaned surfaces of the towers of Westminster Abbey, London, England, are in sharp contrast with the centuries of grime remaining to be cleaned from the rest of the building.

substrate. Although a variety of jet nozzles are available, there is some danger that the salts in the encrusted surface layer or the detergent used will be driven further into the surface.

Adaptations of water-cleaning methods use heated water where the surface is prewet with a gentle mist of water containing a cleaning agent. This pretreatment increases the effectiveness of the method, but is time-consuming and leads to deeper penetration of water and salts into the masonry surface.

Disposal of used water is a problem that must be considered.

4.17.2 Abrasive cleaning

Abrasive cleaning from a jet, either dry or with water, is dirty, is carried through the air to remote areas, and is very destructive to the building's surface. Most of the masonry surfaces have a protective surface layer which, if removed, exposes the surface to even more rapid deterioration. The use of abrasive cleaning is generally not recommended and used only as the last resort.

4.17.3 Chemical cleaning agents

Chemical agents are especially popular for cleaning hard-to-remove dirt, stains, and graffiti. Graffiti has become a serious problem.

Chemical methods of removing heavy deposits of natural stains and dirt include acid cleaners, alkaline cleaners, and organic-solvent cleaners. These are available commercially under numerous trade names. The cleaning chemicals commonly used fall into three classes of compounds: mineral acids, alkalis, and organic solvents.

The acid most commonly used for cleaning masonry surfaces is hydrofluoric acid (HF), the acid that etches glass. It is a weak, little ionized, mineral acid. A dilute water solution is effective in removing deep dirt deposits and stains from the surface of stone, sandstone, granite, brick, and concrete. The acid remaining on the surface can be neutralized with a dilute solution of sodium bicarbonate (baking soda). Hydrofluoric acid is much more effective than water alone, as it removes the deposits with a minimum of mineral surface. It has the disadvantage of not attacking waxy, oily, or paint deposits. In concentrated solutions, hydrofluoric acid is a cancerous and destructive chemical.

Hydrochloric acid (HCl), commonly known as muriatic acid, is a strong, almost fully ionized, mineral acid. It was formerly much used for acid cleaning of masonry, but its use is rapidly being discontinued. Hydrochloric acid has fallen out of favor because its high hydrogen ion concentration etches and dissolves most masonry surfaces. This strong acid dissolves calcareous (calcium-containing) compounds. The chloride ions from hydrochloric acid not only actively promote the corrosion of iron and aluminum, but form soluble chloride salts, such as calcium chloride, which can cause serious efflorescence.

Alkaline cleansers commonly used are sodium hydroxide (NaOH) and ammonium hydroxide (NH_4OH). Potassium hydroxide could be used, but is much more expensive. These chemicals greatly improve the efficiency of cleaning compared to the action of water or steam alone. However, these alkalis require neutralization with acid and a thorough rinsing because of their residual action on the masonry surfaces. The effectiveness of neutralization and rinsing can be monitored by using a pH-indicator paper containing dyes which change color with the pH of the solution. (This test paper is available in chemical supply houses.) To measure the pH, a strip from the roll of test paper is wet with the rinse water. Then its color and pH are matched with the attached color chart. Disposal of the alkaline cleaning solution and the rinsing water can be a problem. In concentrated form these are dangerous chemicals and need to be stored and used with care.

Chemical solvents, needed for the removal of graffiti, are usually applied by hand scrubbing, the use of poultices, or by using paint removers containing a relatively nonvolatile base, which holds down or reduces the volatility of the active solvent component. Many different solvents can be used provided they do not violate pollution emission standards. Chlorinated organic compounds are especially good for the removal of tars, oils, and some paints. Storage and use can be problems because of fire hazards and the danger of breathing the vapors in poorly ventilated storerooms and work spaces.

Because of the technical nature of the stains or dirt on building surfaces, as well as the chemical nature of the masonry surfaces, it is well to turn the job over to professional cleaners who use tested proprietary cleaning compounds.

4.18 Hazardous Chemicals

Hydrofluoric acid is a solution of hydrogen fluoride gas in water. It is toxic by ingestion or inhalation. In concentrated form it is highly corrosive to skin and mucous membranes. Its threshold limit value is 3 parts per million.

Hydrochloric acid, a solution of hydrochloric acid in water, is toxic. It is a strong irritant to the eyes and the skin.

Sodium hydroxide (NaOH) solutions are corrosive and are strong irritants to the skin and, especially, the eyes.

Ammonium hydroxide is a water solution of ammonia (NH_3). The vapor and the solution are extremely irritating to the skin and, especially, the eyes.

Chemists constantly work with acids, alkalis, and flammable liquids. Workers must wear protective clothing and masks or goggles for they know that any such chemical accidentally splashed on clothes or skin should be treated immediately. Modern plants and laboratories have a nearby shower with a chain to pull for showering (clothes and all) using a forceful stream of water. In case of an accident, workers are trained to rush to the shower fully clothed, pull the chain, and flood the chemical with water. Eye fountains are available for flushing the eyes in case of a chemical hitting the eyes. The general rule is: if there is an unexpected liquid in the eyes, rush to the water and rinse first, then think what to do next.

4.19 OSHA Responsibilities for Workers

Safety precautions begin with the most obvious situations. For example, always wear a protective hard hat, safely shoes or boots with steel toes, and face shields or safety goggles if exposure to cement, mortar dust, or flying chips is expected. Dust masks should be worn to protect the lungs.

Especially hazardous are high places in wet or icy weather. Roofs, flashings, scaffolding, ladders, or parapets are always dangerous, particularly in high winds or bad weather. A safety belt should be worn by anyone working on a sloping roof.

4.20 Masonry Problems and Their Prevention

Structural failure does not always mean that the cement must have greater strength; more often failure occurs in the mortar. Failures occur when stresses converging on a weak point in the structure exceed the strength of the material in a flawed region. This flaw can be a fragile foreign particle, a large air bubble, or a poorly mixed or poorly cured region in the mortar. In other words, structural failures usually occur at a much lower stress than the ultimate strength of the material. For example, since bricks or concrete blocks are cured under ideal conditions, flaws are more apt to occur in the mortar or in its bond to the masonry unit. Thus correction would require that the excess mortar be reduced. The structure is only as strong as its weakest link. This principle applies to all types of cementitious adhesives.

4.21 Environmental Hazards

Although stone, baked clay products, and cement are some of the most durable building materials, even they can deteriorate in contact with the chemicals in the soil or desert sand and moisture.

The oxides of nitrogen from automobile exhaust, smog, and vog (volcanic emissions) and the industrial smokestack emissions of oxides of sulfur carried on the smoke particulates all become acids. In the presence of moisture in the clouds, acid rain results. This acid rain not only kills trees, it also reacts with and slowly destroys the surfaces of our buildings.

Terms Introduced

Clay masonry units	Mortar aggregates
Concrete blocks	Mortar ingredients
Glass blocks	Pozzolanic materials
Grout masonry	Stone masonry
Masonry cements	Tile masonry
Masonry cleaning	Travertine
Masonry mortar	Water/cement ratio
Masonry units	Wattle and daub
Mortar admixtures	Wythe

Questions and Problems

4.1. Name the ingredients of a typical portland cement mortar and describe the purpose of each ingredient.

4.2. Describe the differences between type M and type O mortars.

4.3. What is blended cement? Describe its action with water, which gives it cementitious properties.

4.4. What chemicals does the term "lime" designate and why is it used in mortar?

4.5. Why do the aggregates used in mortar and grout need to be graded in size?

4.6. What are the chief ingredients of grout and what is the purpose of each?

4.7. What is a wythe?

4.8. How does the addition of fly ash and slag contribute to the strength of concrete blocks?

4.9. Speculate about the ways in which the use of aluminum powder as an admixture could cause something in the pressurized steam method of making concrete blocks to go wrong.

4.10. Although both clay bricks and tiles are made in heated kilns, they have very different properties. Explain.

4.11. What makes plaster strong and durable?

4.12. Compare the methods for restoring masonry surfaces.

4.13. Describe the health hazards involved in using chemicals to clean masonry surfaces.

4.14. What safety measures are recommended and what immediate treatment should be given if a person is splashed with a chemical?

Suggestions for Further Reading

Christine Beall, *Masonry Design and Detailing for Architects, Engineers, and Contractors,* 3d ed., McGraw-Hill, New York, 1993.

James R. Clifton, ed., *Cleaning Stone and Masonry,* STP 935, American Society for Testing and Materials, Philadelphia, Pa., 1983.

Mark London, *Masonry: How to Care for Old and Historic Brick and Stone,* Preservation Press, Washington, D.C., 1988.

Kenneth J. Nolan, *Mastering Masonry: How to Work with Bricks, Blocks, Concrete and Stone,* Jonathan David, Middle Village, N.Y., 1980.

E. Nonveiller, *Grouting Theory and Practice,* Elsevier, New York, 1989.

Specifications for Concrete Masonry Units, National Concrete Masonry Association, Herndon, Va., June 1987.

Harold B. Olin, *Construction: Principles, Materials and Methods,* Interstate Printers and Publishers, Danville, Ill., 1983.

Chapter 5

Metals

5.1 Introduction	122
5.2 Metals in Construction	123
5.3 Metallic Properties, Bonding, and Structure	123
5.4 Metallic Structures	124
5.4.1 Examples of Metallic Structures	124
5.4.2 Examples of hcp Crystal Structure	125
5.4.3 Examples of fcc Crystal Structure	127
5.4.4 Example of bcc Crystal Structure	129
5.5 Solid-Solution Metallic Structures	129
5.6 Formation of Solid Solutions	130
5.7 Ferrous Metal Structures	131
5.8 Basic Phase Diagrams	132
5.8.1 Simple Eutectic Formation	132
5.8.2 Compound Formation	133
5.8.3 Solid-Solution Formation	133
5.9 Allotropic Forms of Iron	133
5.10 Smelting of Iron Ores	134
5.11 Iron-Carbon Phase Diagram	134
5.12 Iron and Its Alloys	136
5.12.1 Austenite Formation Temperature	137
5.12.2 Ferrite Formation Temperature	137
5.12.3 Cementite Formation Temperature	137
5.12.4 Pearlite Formation Temperature	138
5.12.5 Martensite Formation Temperature	138
5.12.6 Bainite Formation Temperature	138
5.12.7 Stainless-Steel Alloys	139
5.13 Work Hardening	141
5.14 Heat Treatment	141
5.15 Annealing	142
5.15.1 Quenching	142
5.15.2 Tempering	142
5.16 Common Nonferrous Alloys	143
5.16.1 Aluminum-Silicon Alloys	144
5.16.2 Aluminum-Copper Alloys	144
5.16.3 Aluminum-Magnesium Alloys	144
5.16.4 Copper and Its Alloys	145

5.16.5 Copper-Zinc Alloys	145
5.16.6 Copper-Tin Alloys	145
5.16.7 Copper-Nickel Alloys	146
5.16.8 Lead-Tin Alloys	146
5.16.9 High-Performance Metal Composites	146
5.17 Metallic Corrosion	147
5.18 The Corrosion Process	148
5.19 Example of the Oxidation of Aluminum	148
5.20 Galvanic Corrosion between Dissimilar Metals	149
5.20.1 Harnessing Corrosion in a Battery	149
5.20.2 Example of Galvanic Corrosion	150
5.20.3 Example of a Sacrificial Anode	151
5.21 Standard Electrode Potential Series	151
5.22 Local or Nonuniform Corrosion	153
5.23 Corrosion-Resistant Metals and Alloy Steels	155
5.24 Don't Give Up	157
5.25 Balancing Redox Equations	157
5.26 Displacement Series	158
5.27 Prediction of Galvanic Corrosion	158
5.27.1 Example of a Concentration Cell	159
5.27.2 Sacrificial Anode Cell	160
5.28 Surface-Corrosion Protection	160
5.29 Corrosion Protection from Circulating Water	161
5.30 Metal Cladding	161
5.31 Electrodeposition of a Metallic Coating	162
5.32 Galvanic Protection with Sacrificial Anodes	163
5.33 Example of a Zinc Sacrificial Anode	163
5.34 Sacrificial Metallic Pigment Coating	165
5.35 Corrosion Protection with Applied Voltage	165
5.36 Passivation by Anodizing	166
5.37 Organic Protective Prime Coat	166
5.38 Phosphoric Acid Treatment of Metallic Surfaces	167
5.39 Corrosion-Protective Pigments	167
5.40 Chemical Conversion Coatings	168
5.41 Building Problems Related to Metals	169
Terms Introduced	169
Questions and Problems	170
Suggestions for Further Reading	171

5.1 Introduction

It is not surprising that metals are widely used in all types of construction. Aluminum, the most abundant metallic element, makes up an estimated 8 percent of the earth's crust. Iron is in second place with about 5 percent. Metallic iron, alloyed with a small amount of nickel, is found in meteorites that strike the earth. This natural source of iron, when fashioned into tools and weapons, influenced the development of early civilizations. Iron continues to play a vital part in our lives.

The properties of metals, the nature of metallic bonding, and common metallic structures are described in this chapter. Models are used to illustrate the relationships between metallic structures, properties, and performance. Iron and iron alloys are selected as typical metals to illustrate the varied metallic properties. With the help of temperature composition diagrams, alloy formation is more clearly understood. Aluminum, copper, and lead are selected as typical nonferrous metals to relate properties to their structure. The nature of the corrosion process and corrosion prevention are fully investigated. Building problems and the environmental hazards associated with metals are carefully detailed.

5.2 Metals in Construction

Metals can be divided into two types: ferrous and nonferrous. Iron and its many steel alloys are ferrous metals; aluminum, copper, and zinc are some of the common nonferrous metals. Iron, so often used in construction, has been investigated extensively over the years. Therefore, in this book we use iron and the alloys of iron to illustrate many of the physical properties of metals.

The method of preparing iron from its ores results in carbon being a common ingredient in iron and steel. The presence of carbon has a marked effect on the metallic properties of iron and iron alloys. Small amounts of other alloying metals change the properties so much that our study of metals must also include intermetallic compounds and solid solutions. These are more easily visualized with metallic-structure models and the phase diagrams.

5.3 Metallic Properties, Bonding, and Structure

Metals are typically so malleable that they can be hammered into thin sheets and are so ductile that they can be drawn into thin wires. In addition, metals have high electrical conductivity, thermal conductivity, and a silvery metallic luster. Neither ionic bonds with their tightly held electrons nor covalent bonds with their tightly shared electrons can account for such metallic properties. By contrast, the valence electrons which bond the atoms together in metals need to be free to move between the metallic atoms and permit the atoms to glide past one another.

Thus a metallic bond can be thought of as a "cloud" of valence electrons moving freely between the atoms of a metal to produce strong relatively nondirectional bonds. This electron cloud does not restrict the location of the metallic bonds provided the atoms are packed together in an orderly structure. Since the metallic bond is nondirectional, the atoms glide over one another with a minimum of effort. This provides the properties of malleability and ductility. Also, the electron cloud (a sea of valence electrons) is free to flow through the metal, thus providing electrical conductivity. Thermal conductivity results from the transfer of heat (kinetic energy) by the cloud of electrons. And the electron cloud accounts for the highly reflective light that produces the characteristic metallic luster.

There are 12 different geometrical structures possible. However, most fall into one of three types. The most common structure is the body-centered cubic (bcc) one. The other two (the most closely packed) are the hexagonal close-packed (hcp) and the face-centered cubic (fcc) structures. These structures have slight, but important, differences, which are seen easily from models. The structures are basic to an understanding of metals and their alloys.

5.4 Metallic Structures

The purpose of making and using models is to visualize the structures of the more common industrial metals and their alloys more easily. If you are unable to construct the models personally, it will be helpful to picture the operations mentally and closely examine the models in the figures.

How would you build models of the three most common metallic structures? Assume that the atoms are spherical and can be represented by Styrofoam balls.

5.4.1 Examples of metallic structures

To construct models of the three most common metallic structures, you will need 46 Styrofoam balls, each 5.1 centimeters (2 inches) in diameter, a package of toothpicks, and a package of chenille-covered wire (available at floral, hobby, or party supply stores).

Start by arranging in a box lid or tray 15 of the balls in three orderly, closely packed rows to represent what a layer of metallic atoms might look like. Assemble the balls into three rows (of five balls each) using toothpicks to hold them in rows. Remember that metallic bonding allows atoms to glide past one another to fit as closely together as possible.

Metals 125

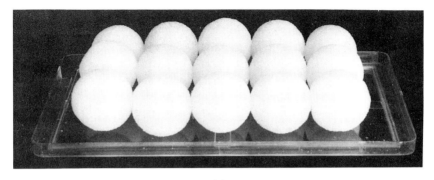

(a)

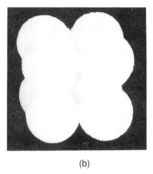

(b)

Figure 5.1 (*a*) Model layer of a metallic crystal. (*b*) Model of cubic unit cell.

It will soon be apparent that there are two ways to arrange the balls in an orderly symmetrical pattern. The first, most orderly, way to line up the balls into rows and columns is shown in Fig. 5.1*a*. If this bottom layer is extended upward with identical layers of balls lined up on top of one another, the metallic structure would be that of a simple cubic crystal. However, you will notice that the balls in this typical layer of a cubic structure are not as closely packed as possible.

In practice, metals with this cubic structure are readily stress-hardened (work-hardened) to form a more stable closely packed structure. A unit cell (simplest repeated atomic unit) is often used to visualize a crystal type. A simple cubic unit cell is shown in Fig. 5.1*b*.

5.4.2 Examples of hcp crystal structure

Rearrange the rows of balls in the tray by sliding the middle row sideways a distance of half a ball. This arrangement, with alternating

(a)

(b)

Figure 5.2 (a) Model of hcp atomic layer. (b) Model of hcp unit cell.

rows of balls fitted into the hollows formed between balls in adjacent rows (shown in Fig. 5.2a), is typical of the atomic layers in the close-packed crystal structures. (This is the structure to use to construct hcp and fcc models.) To hold the balls in place, it is advisable to attach the rows to each other. Notice that patterns in the layer of close-packed balls can be picked out in the shape of hexagons and triangles. With a felt-tip pen, mark the position of a hexagon at one end of the basic 15-ball layer, as shown in Fig. 5.2a.

Now with the other balls and chenille wire, prepare two units (arrangements) of balls: (1) connect three balls together with 1-inch lengths of chenille-covered wire to make the shape of a triangle; and (2) connect six more balls around a seventh ball to form a hexagon. (The chenille wire is used so that balls can be more easily squeezed into place.)

Assemble the hcp model with these two units and the three rows of ink-marked balls in the tray. To start, form a second layer by placing the triangular unit of balls in the holes formed inside the ink-marked balls. Then make a third layer by fitting the hexagonal-shaped ball unit on top of the triangular unit. Rotate the top hexagon, if necessary, so that each of its balls is lined up directly above a similar ball of the hexagon in the bottom layer. This completes the hcp structure shown in Fig. 5.2*b*. This model could be continued upward indefinitely, with a triangular unit for a fourth layer and a hexagonal unit lined up, as before, as a fifth layer. If the hexagonal units are designated by the letter *A* and the triangular units by *B,* the hcp structure (as viewed from the side) can be identified by its *ABAB* repeating structure.

Notice that the central ball (atom) in the center of the hexagonal model has not only the six atoms of the hexagonal unit touching it, but all three of the atoms in the triangular units that are below and above it. The number of nearest neighbors of an atom is called the coordination number. For the hcp structure, the coordination number is seen to be 12. This is the largest number of identical close neighbors possible in a metallic structure.

5.4.3 Examples of fcc crystal structure

The hcp model can be easily converted into a model of a close-packed fcc structure. Just rotate the third layer of balls through an angle of 60°. Each ball in the upper hexagon layer is now over a space between balls. This completes the fcc model, as shown in Fig. 5.3*a*. You will note from this figure that the balls in the third layer are no longer directly over those in the hexagon in the bottom layer. As viewed from the side, the top hexagonal layer resembles neither the *A* nor the *B* shape, so it is designated as *C*. This close-packed fcc structure, if repeated upward, would be identified as a repeated *ABCABC*. It can be seen that its coordination number is also 12.

This close-packed fcc structure would need to be expanded by fitting in many adjacent balls before it would be apparent that it is a face-centered cube.

It is easier to visualize this fcc structure by constructing the fcc model called a unit cell. To do this, assemble four balls into a square so that the balls are touching. Use short pieces of chenille-covered wire as before. Make two more identical squares. Now assemble the fcc structure by placing the three squares on top of one another, but

(a)

(b)

Figure 5.3 (a) Model of fcc atomic arrangement. (b) Model of fcc unit cell.

rotate the middle square by an angle of 60° so that it is no longer lined up with the balls above and below. Fit the middle layer snugly (close-packed) into the hollows between the upper and lower sets of balls. This completes a second model of the fcc metallic structure, shown in Fig. 5.3b.

Figure 5.4 Model of bcc unit cell.

5.4.4 Example of bcc crystal structure

A bcc structure can be most easily seen with the bcc unit cell model. First construct two more square units of four balls each from the remaining balls. By placing a single ball in the center between these two symmetrically oriented units, a bcc model is formed, as shown in Fig. 5.4.

Note that the ball in the center of this body-centered cube has eight nearest neighbors. Each iron atom in a bcc crystal is at the center of a cube of eight other iron atoms. With a coordination number of 8 atoms it is apparent that the bcc structure is not quite as closely packed as the hcp or the fcc structures with their coordination numbers of 12. Note that each of these three structures has small open spaces between the metallic atoms into which smaller atoms, such as carbon, might fit. The following section explains where the carbon comes from.

5.5 Solid-Solution Metallic Structures

Even the close-packed metallic structures contain empty spaces (holes), assuming atoms are spherical. For example, in the fcc and hcp metallic structures, about 26 percent of the volume is empty. Nearly 29 percent of the bcc volume is empty. This empty space is made up of a number of holes. In these three close-packed structures each atom is surrounded by 24 triagonal holes, 8 tetrahedral holes, and 6 octahedral holes. If the holes are large enough, other atoms can slip into a hole to form what is called an interstitial solid solution. This is not a compound because only part of the holes may be occupied, resulting in a variable composition. By contrast, a chemical compound such as cementite (Fe_3C) has a fixed carbon content of 6.6897 percent.

The holes in gamma iron are nearly half the diameter of the carbon atom, causing the solubility of carbon to be practically zero in austenite iron. However, the holes in alpha iron are comparable to the size of the carbon atom, allowing an interstitial solubility of about 2 percent carbon in austenite iron.

5.6 Formation of Solid Solutions

In a solution it is customary for the term "solvent" to be used for the atom that is in excess. The atom in the minority is called the solute. This also applies to an ion or compound if either is in a water solution. Since the effective diameter of the solute particle changes with changes in temperature, the maximum solubility of one element in another usually decreases as the temperature drops.

An interstitial solid solution of one element in another requires that the space (holes) between solvent atoms be large enough to accommodate the solute atom. Substitutional solid solution is the other type of metallic solid solution. This results from the replacement (substitution) of solute atoms for solvent atoms in the solidified crystalline structure. Note that a solid solution would occur more slowly if both elements are solid than if they have more freedom of movement in the molten condition. For example, semiconductors (which made solid-state electronics possible) result from the diffusion (doping) of gallium atoms into the crystal of pure solid silicon to form a solid solution. Because of the nature of metallic bonding, neither the intersubstitutional nor the substitutional bonding can be classified as a "normal" chemical compound.

The Hume-Rothery rule summarizes the requirements for the formation of a substitutional alloy between two elements: the two elements should have atomic diameters within 15 percent of one another; the elements should have the same type of crystals, such as fcc or bcc; and they must not be chemically reactive. Table 5.1 lists the effective atomic diameters and the crystal types of some common elements for use with the Hume-Rothery rule. The third criterion for solid-solution formation is given for the elements with similar chemical reactivity, as shown by their position in the electrochemical series and displacement series treated later in the chapter.

TABLE 5.1 Solid-Solution Atomic Diameters

Element	Symbol	Crystal structure	Atomic diameter, Å	Melting point, °C
Aluminum	Al	fcc	2.86	660
Cadmium	Cd	hcp	2.97	321
Carbon*	C	Hexagonal	1.54	—
Chromium	Cr	bcc	2.48	1875
Copper	Cu	fcc	2.56	1083
Iron	Fe	bcc	2.48	1536
	Fe	fcc	2.54	1400†
Lead	Pb	fcc	3.50	327
Magnesium	Mg	hcp	3.21	651
Manganese	Mn	Cubic	2.24	1245
Molybdenum	Mo	bcc	2.72	2610
Nickel	Ni	fcc	2.49	1453
Silicon	Si	Diamond	2.35	1410
Silver	Ag	fcc	2.88	961
Sulfur	S	Orthorhombic	2.12	119
Tin	Sn	Tetrahedral	3.02	2996
Titanium	Ti	hcp	2.94	1668
	Ti	bcc	2.89	—
Zinc	Zn	hcp	2.66	420

*Graphite.
†Transition formation.

5.7 Ferrous Metal Structures

Iron, although a pure element, occurs in four different allotropic structures (different crystalline forms of the same element) called alpha, beta, delta, and gamma iron. Neither the beta nor the delta forms of iron are commercially important. The delta form, with its bcc structure, is stable over such a narrow high-temperature range of 1538 to 1394°C (2800 to 2541°F) that it is of little commercial use. The two common commercial forms are gamma iron, with its fcc structure, which is formed at temperatures from 1394 to 912°C (2541 to 1674°F), and alpha iron, which has a bcc structure and is formed from 912 to 273°C (1674 to 460°F).

When gamma iron contains carbon, it is called austenitic; when alpha iron contains carbon, it is called ferritic. Other steel alloys are called austenitic if they have the same gamma structures. Also, it is customary to call steel alloys ferritic if they have the alpha structure.

Alloys can be identified in two ways. They can be etched with acid to expose their crystal grain boundaries, and they can be examined under a metallographic microscope (viewed with top lighting). Names given to identify the alloy, such as austenite, were given to several

5.8 Basic Phase Diagrams

When two metallic elements are heated to form a molten mixture and slowly cooled, the alloy usually has very different physical properties than either of the two metals from which it was made. Phase diagrams show the makeup of the stable mixtures that exist at each temperature and composition at equilibrium. These diagrams are needed to appreciate common metals and their alloys.

The three basic types of phase diagrams are (1) a formation of a simple eutectic with a lower melting point than either metal; (2) the formation of a metallic compound that may have either a higher or a lower melting point; and (3) a solid solution in which one metal dissolves in the other solid in unlimited proportions.

5.8.1 Simple eutectic formation

Figure 5.5 shows a temperature composition phase diagram under equilibrium conditions of two pure metals, A and B, which form a eutectic. A eutectic is a mixture of fine crystals of A and of B. For example, bismuth melts at 271°C (520°F) and cadmium at 321°C (610°F) to form a eutectic that melts at 144°C (291°F). Solders and brazing rods make use of low eutectic melting points.

Figure 5.5 shows that compositions at temperatures enclosed in region 1 consist of melted A and melted B. Region 2 has solid A and

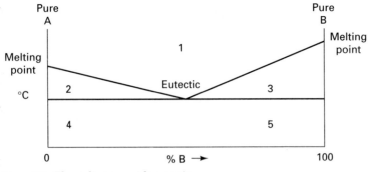

Figure 5.5 Phase diagram with eutectic.

melted *A* and *B*. Region 3 has solid *B* and melted *A* and *B*. Region 4 has solid *A* and solid eutectic. Region 5 has solid *B* and solid eutectic.

5.8.2 Compound formation

Metallic compounds act like pure metals in phase diagrams. Their phase diagrams look like two eutectic diagrams placed together with their adjacent vertical axes combined. The melting point of the compound may be higher or lower than that of the pure metals *A* and *B*. An illustration would be the compound formed between iron and carbon, iron carbide (Fe_3C).

5.8.3 Solid-solution formation

Figure 5.6 is the phase diagram for two metals that form a solid-solution alloy of one metal in all proportions. Similarly sized atoms allow each to crystallize in the lattice of the other to make a solid solution. Thus their physical properties, such as density, tensile strength, and electrical resistivity, change continuously from that of one metal to that of the other. Examples are nickel-copper and iron-carbon alloys.

Figure 5.6 shows that region 1 has only melted *A* and *B*. Region 2 has a solid solution of *A* and *B* and molten *A* and *B*. Region 3 has a frozen solid solution consisting of varying amounts of *A* and *B*.

5.9 Allotropic Forms of Iron

As the metal most important to the building industry, iron has been investigated extensively. It is thus appropriate to use iron and its alloys to illustrate the unique properties of metals.

Many useful kinds of carbon-steel alloys are formed by the alpha and gamma allotropic forms (several crystalline structural forms of an element) of iron. The gamma form of iron, which is stable at a

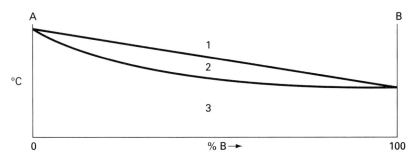

Figure 5.6 Phase diagram with solid solution.

higher temperature, is an fcc crystal. This means that an atom of iron is at each of the eight corners of a cube, with another atom of iron embedded in each of the six faces of the cube. The alpha allotropic iron crystallizes in a body-centered cube. Besides the atoms of iron at each corner, it has an iron atom in the center of the cube. These allotropic forms of iron have differently sized spaces between iron atoms for the carbon atoms to fit, or dissolve, into the crystal lattice. The number of available spaces limits the amount of carbon that can dissolve in the solid solutions.

5.10 Smelting of Iron Ores

Iron ore is mined from many countries as iron oxide and iron carbonate. The oxides are reduced to metallic iron and carbon monoxide by heating with coke or some other form of carbon. The carbonate ore is decomposed to the oxide by heating. However, the reaction between two solids, such as iron oxide and coke, is too slow to be commercially profitable. The blast-furnace process for producing iron from its oxides converts coke to carbon monoxide to take advantage of the increased rate of reaction when one of the reactants is a gas.

In this blast-furnace process, powdered iron ore, coal, and flux are fed into the top of the furnace and heated air is forced in at the bottom. Burning coke produces carbon monoxide and the heat needed to maintain the reaction temperature. A flux is needed to convert the excess acidic impurities in the ore, silicon dioxide (in the sand and clay) or calcium oxide (from calcium carbonate), to an easily melted slag. Just enough flux is used to convert any excess acidic or basic impurity to additional slag. As the iron is formed, the molten metal falls to the bottom of the furnace, where it is drawn off in a continuous process. The slag floats on top of the molten iron and is drawn off for sale as an ingredient of cement. Any unreacted coke carbon dissolves in the molten iron, or if the amount of carbon reaches 6.7 percent, the carbon reacts with iron to form the compound called cementite (Fe_3C), which is drawn off with the molten iron.

5.11 Iron-Carbon Phase Diagram

Figure 5.7 shows part of the phase diagram of iron and carbon from 0 percent carbon to the formation of the Fe_3C compound. It is actually the phase diagram of iron cementite since iron is one vertical axis and pure cementite is the other. This diagram shows the range of stability of four allotropic forms of iron, two solid solutions, and the compound cementite. The diagram also includes the carbon alloys known as wrought

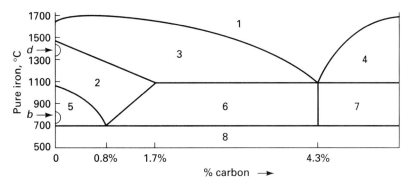

Figure 5.7 Basic iron-carbon phase diagram.

iron, steel, and cast iron. This can be thought of as a phase diagram of iron versus iron carbide since it only extends to the iron carbide formation. This region of the diagram includes alpha, beta, gamma, and delta—the four allotropic forms of iron (different crystalline forms of the element). Only the alpha and gamma forms are of industrial significance and are discussed in detail. It is evident that the diagram includes each of the features of the two previous phase diagrams.

Iron and carbon form the compound cementite (Fe_3C), which contains 6.67 percent carbon. A eutectic of iron and cementite is formed with 4.3 percent carbon at 1448°C (2638°F) and a eutectoid (solid-solution transformation) containing 0.8 percent carbon at 727°C (1341°F).

Austenite, found in region 2 of Fig. 5.7, is a solid solution of carbon in gamma iron (an fcc crystal form of iron). Ferrite is a solid solution of carbon in alpha iron (a bcc crystal). Cementite was the name given to the chemical compound iron carbide.

Region 1 of Fig. 5.7 begins with carbon dissolved in the molten iron. Region 2 consists of solid austenite and a molten mixture of carbon and iron. Region 3 has solid cementite and a molten mixture of carbon and iron. Region 4 consists of solid austenite. Region 5 has solid austenite and solid ferrite. Region 6 has two solids, austenite and a solid made up of alternating layers of solid austenite and cementite. In region 7 there is solid austenite and solid cementite. Finally, region 8 consists of two solids, ferrite (alpha) and cementite [iron carbide (Fe_3C)].

Notice that the transition from region 2 to region 4 is caused by the change of the gamma form of iron in the austenite solid solution to the alpha form of iron in ferrite that is stable at the lower temperature. Since alpha iron cannot hold as much carbon in solid solution as gamma iron, the excess carbon reacts with some of the iron to form iron carbide.

Neither the beta nor the delta forms of iron are of any industrial importance. Beta iron, with a bcc structure, is found in region b. Another allotropic form of iron, delta, also with a bcc structure, is found in region d. Delta is stable at very small percentages of carbon and over a temperature range of 1538 to 1394°C (2800 to 2541°F). The maximum solubility of carbon in the solid solution of delta iron is 0.10 percent. Delta iron is not commercially useful, and its formation is to be avoided during welding or other treatment involving its melting. Fortunately hot-working and other processes do not reach such a high temperature; most equipment would not stand this temperature.

Gamma iron has an fcc structure. It is the stable form of iron over the temperature range of 1394 to 912°C (2541 to 1674°F). This type of iron forms an interstitial solid solution of carbon, called austenite, that can dissolve as much as 2 percent carbon.

Alpha iron has a bcc structure like delta iron. However, existing at a much lower temperature, the holes in its metallic structure are smaller due to the smaller thermal vibrational motion of the iron atoms.

Wrought iron, with a low carbon content (about 0.035 to 2.00 percent carbon) and a trace of manganese and phosphorus, also contains from 1 to 4 percent slag dispersed in the alloy. The presence of slag makes wrought iron sufficiently ductile to be cold-worked.

Cast iron is defined as iron containing more than 2 percent carbon, but usually not more than 4.3 percent. Since the maximum interstitial solubility is 2 percent, the extra carbon is in the form of cementite (iron carbide). The presence of iron carbide makes the cast iron hard, brittle, and difficult to shape mechanically. Therefore it has to be cast in a mold. Cast iron is usually manufactured by melting scrap and cast iron.

5.12 Iron and Its Alloys

The two carbon steels austenite and ferrite have quite different physical properties, such as ductility, strength, and corrosion resistance. Because of the sluggish movement of atoms through a solid, the rearrangement of the atoms into stable forms may take years at room temperature. However, if their temperature and thus the kinetic energy of the atoms are increased during heat treatment, the time is greatly reduced.

Iron carbide has an orthorhombic structure. The unit cell has 12 iron (Fe) atoms and 4 carbons (interstitial) in the closed-packed structure.

The iron-carbon phase diagram in Fig. 5.7 shows the materials as they exist in equilibrium in the temperature ranges shown. For example, annealing cementite decomposes it into iron and graphite (carbon).

The room-temperature composition of the eutectic formed at 0.77°C (33°F) is 88 percent ferrite and 12 percent cementite. This structure (layered), known as pearlite, contains 61 percent ferrite and 39 percent pearlite.

Many different kinds of steels can be obtained by the eutectoid decomposition from fcc to bcc since ferrite, cementite, pearlite, martensite, and bainite can be obtained by controlling the cooling rate and the amount of carbon and other alloying elements present.

5.12.1 Austenite formation temperature

Austenite, an alloy of carbon in iron, is a solid solution of carbon in gamma iron (fcc). The maximum solubility of carbon in this fcc iron structure is 2 percent. The carbon atoms fit into the holes between the iron atoms to form an interstitial solid solution. Austenite, ductile enough to be cold-worked to increase its hardness, is nonmagnetic and has a larger electrical resistance and thermal expansion coefficient than ferrite.

5.12.2 Ferrite formation temperature

Ferrite is a solid solution of carbon in the bcc structure of alpha iron. Although the holes in this bcc alpha structure are more numerous than in the fcc austenite structure, they are too small to accommodate carbon atoms. A few carbon atoms are able to fit into the holes in the edges of the structure. This reduces the interstitial solubility of carbon in ferrite to 0.025 percent. Note that carbon is too small to replace iron atoms to form a substitutional solid solution since the diameter of the carbon atom is 40.5 percent larger than that of iron. Ferrite is more ductile than austenite and is easily cold-worked. Chromium, molybdenum, tungsten, and silicon, by substitution, form alloys that are much used in the building industry.

5.12.3 Cementite formation temperature

Cementite (iron carbide) has the formula Fe_3C and is 6.68 percent carbon. Like other metallic carbides, such as the abrasive carborundum (silicon carbide), cementite is a very hard alloy. Cementite is ferromagnetic below 2300°C (4172°F). If it is annealed or hot-worked, cementite slowly decomposes into iron and carbon.

5.12.4 Pearlite formation temperature

Pearlite consists of alternating platelets of ferrite and cementite. The pearllike luster of the layers can be seen with a metallographic microscope. This layered formation gives pearlite the lustrous appearance of mother-of-pearl, from which it is named.

In the austenite eutectoid (0.77 percent carbon formed at 727°C in steel) pearlite forms on slow equilibrium cooling below the critical temperature of 727°C (1341°F). A layer of ferrite grows on either face of a cementite nucleus, which forms the intermediate layer. Pearlite has alternate platelet layers of ferrite and cementite.

The formation of pearlite involves two steps: (1) the diffusion of carbon and its distribution in two layers (ferrite and cementite) and (2) the transformation of the austenite structure (Fe) to the orthorhombic cementite (Fe_3) and the ferrite (bcc). The transformation involves an energy release of heat with undercooling, but this drop in temperature decreases the rate of diffusion of carbon to the growth site. There is an optimum temperature in between. If the temperature is held at 727°C (1341°F), the result is 100 percent austenite. If allowed to cool in the insulated furnace overnight, thick platelets of pearlite form; if it cools a little faster in air, the result will be thin platelets.

5.12.5 Martensite formation temperature

Martensite carbon steel is formed when austenite is quenched. Martensite is metastable and, unless heat-treated to speed up the conversion to stable ferrite and iron carbide, the metastable state may exist for years at ambient temperature. Since some regions of the phase diagram represent a metal more easily machined than others, this change may be obtained intentionally. For example, martensitic steels are some of the hardest steels, but they have low ductility. This is a common tradeoff with metals. Martensite steel is too brittle (when formed by quenching) to be of value in industry. Fortunately, it can be tempered (reheated to a specified temperature) to increase its toughness without sacrificing too much hardness.

5.12.6 Bainite formation temperature

Bainite is similar to pearlite since it is formed by slow cooling to allow diffusion of carbon to take place. However, it consists of an intimate mixture of crystals rather than the platelets in pearlite. Pearlite formation is started by cementite precipitation, whereas bainite is started by ferrite formation. Bainite is formed at a temperature below 727°C (1341°F), the critical temperature needed for pearlite forma-

Figure 5.8 Extensive use of steel: workers assemble wire and reinforcing bars for floors, beams, and columns in preparation for pouring concrete.

tion. In the case of AISI 1080 steel (American Iron and Steel Institute), the best temperature for bainite formation is about 540°C (1004°F). Figure 5.8 shows the use of iron alloys.

5.12.7 Stainless-steel alloys

There are three main classes of stainless steels: austenite, ferrite, and martensite. The stainless-steel alloys differ in composition, metallurgical structure, workability, magnetic nature, and corrosion resistance. They contain as much as 25 percent alloying elements, which add to the cost of corrosion-protected alloys. Each of the alloys contains chromium, which protects the surface from corrosion by forming a thin, tightly adherent transparent film of chromium oxide. The carbon content must be kept low to prevent the chromium from being removed from the alloy in the form of chromium carbide. The formation not only means less chromium content, but allows precipitation of the chromium carbide on the grain boundaries to form intergranular corrosion. One method of corrosion prevention is to modify austenitic stainless steel by the addition of a very small amount of titanium or aluminum, which forms carbides more readily than chromium.

Figure 5.9 Frederick R. Weisman Art Museum, University of Minnesota, Minneapolis, covered with a brushed stainless-steel facade resembling exploded popcorn.

Austenitic stainless steels of the 18-8 type (percentage of chromium and nickel, respectively) are most often used when highly corrosion-resistant decorative and nonmagnetic properties are required (Fig. 5.9). They are the most ductile of the stainless steels and are used to form tubes and sheets. The addition of nickel extends the stability of austenitic steel over a much larger region in the phase diagram. This allows many of the alloys to be quenched and work-hardened without the loss of desirable mechanical properties or corrosion resistance. The gamma iron structure makes them nonmagnetic.

Ferritic stainless steels contain chromium but no nickel. This class of steel is produced from an austenite-type steel by slow cooling in order to obtain the alpha form of iron that is stable at ambient temperatures. They are magnetic, and can be hardened somewhat by cold-working, but are not hardened by heat treatment. Their excellent corrosion resistance explains their use in chemical plants.

Martensitic stainless steels are formed by the addition of chromium and can have up to 1 percent carbon. They are precipitation-hardened steels. Both heat treatment and quenching are used in combination with a mechanically applied stress, such as rolling.

Corrosion of stainless steels can occur to a minor extent due to the formation of galvanic cells between ferritic and austenitic crystalline regions. Also, improper heat treatment or excessive carbon content can result in corrosion along grain boundaries. The tendency for intergranular corrosion is less if the carbon content of the steel is low due to the depletion of chromium below the critical amount by its reaction with carbon to form chromium carbide. At high temperatures the more rapid diffusion of carbon favors formation of the carbide and depletion of chromium.

5.13 Work Hardening

Work (or strain) hardening occurs when a metal is rolled, drawn, extruded, or flexed to produce dislocations in the crystal as the atoms are forced to glide over one another. As the atomic planes glide over one another, the resulting dislocations become entangled. As these entanglements increase, greater stress will be required for additional deformation. This increases the strength and stiffness of a metal by the process called work hardening.

5.14 Heat Treatment

Phase diagrams show that an alloy may exist in several different crystal forms. These forms are metastable at room temperature if rapidly cooled or quenched. This permits the choice of utilizing the alloy in one of the metastable forms or in the stable condition at room temperature. The rate of cooling determines the nature of the alloy that results.

Reaction in the solid state is extremely slow. Reactions go faster when heat is applied. Flexing, extrusion, and rolling of metal under load raise the temperature of the metal and increase the speed with which metastable conditions change to more stable microstructures. This is easily demonstrated by continuously bending a wire at one point. The wire becomes hot at that flex point and usually becomes so brittle that it breaks into two pieces. Note that this is a mild form of heat treatment.

Precipitation hardening takes place when the alloy is heated to a temperature at which a desired structure occurs. The crystal-lattice metallic structure is usually interstitial rather than substitutional. Atoms may be squeezed out of the structure; therefore it is called precipitation hardening.

5.15 Annealing

Annealing is an important part of heat treatment. High stakes are involved—the annealing must be done properly and without failure. The annealing temperature needed for the process to occur, the length of time required for the heating, and the final rate of cooling depend on the type of alloy used.

The purpose of annealing may be to relieve stress or to improve ductility and toughness, or it may be to reduce the size of the crystal grains or to change the magnetic properties or the electrical resistance of the material. Strain-relieving annealing and full annealing are typical annealing processes.

Strain-relieving annealing is used to remove internal distortions and dislocations from the structure caused by previous operations. In this treatment the alloy is heated to a predetermined temperature for a specified length of time to restore the more orderly (ductile) structure. It is then slowly cooled to ambient temperature. It can now be machined or shaped. This is often followed by a mild heat-hardening treatment to increase the durability.

In full annealing the alloy is heated to a temperature at which the desired solid-solution structure is stable. The alloy is held at this temperature long enough to allow diffusion of its atoms to form a stable metallic structure. This is followed by slow cooling. In practice the cooling is done in the same furnace with the heat turned off. This slow cooling, at an elevated temperature, allows the alloy to be near enough to equilibrium to form the structures that are stable at that temperature. For example, if a specimen is heated to the temperature at which austenite is stable, then furnace-cooled restoration to the desired ferrite-pearlite structure results.

AISI 1080 steel contains 0.74 to 0.89 percent manganese, which makes the steel considerably more expensive. This steel is more easily annealed because of its longer annealing temperature range.

5.15.1 Quenching

Quenching is the rapid cooling of a hot steel without allowing time for it to complete its transition to the chemical or crystal form which is stable at equilibrium. This can be done by plunging the hot alloy into water or oil.

5.15.2 Tempering

Tempering is used to restore some of the toughness and ductility to steel with as little loss of hardness as possible. The process involves heating to a prescribed temperature to cause partial transition to another more desirable structure, followed by slow cooling. For example, martensite is obtained by quenching a solid solution of iron carbide in austenite. Martensite is strong and hard, but it is too brittle to be of much use commercially. It can be tempered by reheating to the austenite region of the phase diagram for a short specified time. Then, with slow cooling, it again forms a metastable structure of the alloy, which is now tough with little loss in its hardness.

5.16 Common Nonferrous Alloys

Aluminum metal does not occur in nature. It is made by the electrolysis of a molten mixture of bauxite (aluminum oxide ore) and cryolite (sodium aluminum fluoride). In addition to its light weight, it has moderate resistance to corrosion due to the rapid formation of a thin, transparent, tightly adherent aluminum oxide coating. Thicker oxide coatings are produced by an electrolytic process known as anodizing (Fig. 5.10). Aluminum has the fcc structure and forms alloys with copper, magnesium, and silicon (a nonmetal).

Figure 5.10 The anodized marine aluminum alloy dome spans 320 feet. The clear span supports an all-aluminum grid and catwalk system suspended from the dome roof over the arena floor. The lightweight aluminum exerts less stress on the ring tension beam and foundations.

Since cubic structures are more easily stress-worked (cold-worked) than most other structures, aluminum is malleable and ductile. This is in contrast to the lighter magnesium alloys which have a tetragonal (hcp) structure and are not as malleable. These aluminum alloys may contain, in addition, a small percentage of one or more of the metallic elements such as bismuth, chromium, manganese, or lead. These alloys may be used for casting, heat treatment, or stress working. The properties of these alloys vary in strength, machinability, workability, and in their resistance to corrosion. Aluminum foil is a good illustration of the malleability of aluminum.

The large amount of electric energy required for the production of aluminum makes it a prime candidate for recycling.

5.16.1 Aluminum-silicon alloys

Aluminum and silicon form a simple eutectic at 12.6 percent aluminum. Their phase diagram is the simple eutectic type shown in Fig. 5.5. The most useful alloys are on the high-aluminum-content side of the diagram, although it is of interest that the high-silicon side is important in the semiconductor doping (diffusion at elevated temperature) of silicon with aluminum. The solubility of silicon in aluminum decreases from 1.5 percent at 582°C (1080°F) to less than 0.1 percent at room temperature.

5.16.2 Aluminum-copper alloys

An aluminum-copper phase diagram would be more complicated than the iron-carbon phase diagram shown in Fig. 5.7. Aluminum and copper both have the fcc structure. Their atomic diameters differ by only 2.8 percent. The solubility of copper in aluminum drops markedly with the temperature from 5.7 percent at 549°C (1020°F) to less than 0.3 percent at room temperature. A eutectic is formed at about 33.2 percent aluminum.

5.16.3 Aluminum-magnesium alloys

Aluminum and magnesium have a phase diagram like that of iron and carbon, with solid solutions and a eutectic. The solubility of magnesium in aluminum ranges from 15 percent at 452°C (845°F) to about 2 percent at room temperature. This makes it suitable for precipitation hardness.

5.16.4 Copper and its alloys

Copper is used extensively in the pure state and in alloys. In its purest form it is used as a conductor of electricity in the electric wiring of homes and buildings. Only silver is a better electrical conductor than copper. During World War II a limited amount of silver was made into electric wire because of the wartime scarcity of copper. The use of silver is limited due to expense. Although other elements alloyed with copper increase its resistance, the presence of oxygen— up to a trace of 0.05 percent as cuprous oxide (Cu_2O)—increases the electrical conductivity of copper by about 1 percent.

As an alloy, copper is often used for roof gutters and water pipes because of its excellent resistance to corrosion. However, with age copper gutters, flashings, and roofs form a protective brown cuprous oxide coating which, in time, is converted by the moisture and carbon dioxide in the air to an attractive green patina of basic copper carbonate [malachite, $CuCo_3 \cdot Cu(OH)_2$]. Serious rapid corrosion results if iron water pipes and copper water pipes are in direct contact. They must be electrically insulated to avoid contacting each other.

Copper is very ductile and can be drawn into wires and extruded into tubing. It can be cold-worked to increase its strength. However, as is the practice with aluminum, it is usually alloyed with numerous other elements if more strength is needed.

5.16.5 Copper-zinc alloys

Copper-zinc alloys are known as brasses. Although copper is fcc and zinc has the hcp structure, their atomic diameters differ by only 2.8 percent. This allows them to form a limited substitutional solid solution. One solid solution with up to 20 percent zinc is known as red brass. Solid solutions of 20 to 36 percent zinc have a yellow color, are known as yellow brass, and are recognized as the alloy used to make shell cartridges.

5.16.6 Copper-tin alloys

True bronzes are another well-known class of copper alloys. Although copper is fcc and tin has a tetrahedral structure, their atomic diameters differ by a marginal 15.28 percent. The solubility of tin in copper drops dramatically as it is cooled very slowly. Rapid cooling retains the tin in solution as a metastable solid solution. The presence of tin contributes more corrosion resistance, hardness, and abrasion (wear) resistance than the softer brass alloys of copper-zinc.

5.16.7 Copper-nickel alloys

Copper and nickel form a variety of alloys that include Monel, a highly corrosion-resistant alloy. Both copper and nickel form fcc structures. Their phase diagram is similar to that of Fig. 5.6, in which the solid solution is present. The two metals are soluble in one another in all proportions. This total solid solution results from the close fit in diameters of the atoms (2.56 angstroms for copper and 2.49 angstroms for nickel), making it possible for one atom to easily substitute for the other in the fcc structure.

5.16.8 Lead-tin alloys

Lead-tin alloys are interesting because they include the much used lead solder. Lead has the fcc structure. Tin has two allotropic forms: the familiar white metallic form which is tetrahedral and stable from 273 to 18°C (523 to 64°F), and a gray nonmetallic allotropic powder. In very cold climates antique tin artifacts in museums are known to crumble into a gray powder, thought by some people to be a mysterious process called museum disease.

The tin-lead temperature-composition diagram is nearly as complex as that of iron and carbon. Although the melting point of lead is 327°C (621°F) and that of tin is 232°C (450°F), the eutectic composition of 38 percent lead has a low melting point of 183°C (361°F). Lead solder usually contains 50 percent lead instead of the somewhat lower eutectic. Eutectic compositions characteristically melt to form a fluid at a constant temperature. It would be inconvenient for solder to melt suddenly and flow away from the metals to be soldered. This would happen if the eutectic were to be used. Therefore a solid solution slightly removed from the eutectic composition is used so that it can remain a viscous, partially melted solder when heated.

5.16.9 High-performance metal composites

Additional strength can be obtained in an alloy by forming a high-performance fiber-reinforced composite. Fibers such as graphite, silicon carbide, silicon nitride, boron nitride, and alumina are common. The problems frequently encountered in forming such a fiber composite in a molten metallic matrix involve their mechanical and chemical incompatibility.

The anticipated mechanical properties, such as improved strength, toughness, and creep resistance, require a thorough understanding of the transverse and shear fiber-matrix properties. A mismatch results in the matrix cracking and a breakdown of the fiber-matrix interface.

The need of the aerospace industry to find lighter and stronger metallic systems has spurred research and produced exciting results.

Strong lightweight alloys and numerous high-strength fibers are available. Much progress is made in the design of high-performance composites in relatively brittle metallic and ceramic matrices. The chemical reaction between the fiber and the matrix to form an alloy seriously depletes and weakens the fiber. It can form an alloy with mechanical properties incompatible with the matrix.

In the case of a silicon carbide fiber reinforcement of an aluminum alloy, silicon is extracted from the fiber to form aluminum silicide. By keeping the silicon concentration in the matrix above a critical level, the need of the matrix to leach additional silicon from the fiber is relieved.

A more general method is to prevent an element in the fiber from forming an alloy with the matrix by giving the fiber a protective coating. One method is to provide a "sacrificial" coating on the fiber. The fiber can be coated with silicon carbide, which is slowly sacrificed by a reaction with the aluminum-alloy matrix to form aluminum silicide (Al_4Si_3). Another technique is to coat the fiber with alumina, which is chemically inert. Proprietary processes are available, such as the Duralcan molten metal mixing method, which produces low-cost composites. This process permits using conventional casting and fabrication practices.

5.17 Metallic Corrosion

When exposed to moisture and oxygen in the air, metals usually corrode by forming an oxide coating. Iron forms a crumbly reddish-brown oxide coating known as rust. In contrast, some metals, such as aluminum, form a tightly adherent oxide coating. This coating quickly covers the surface and prevents the penetration of oxygen and moisture to the surface. Because this natural oxide film is very thin, colorless, and nearly transparent, it usually goes unnoticed. This spontaneous reaction should not be surprising because aluminum oxide (not the element aluminum) is found in nature as the mineral bauxite (aluminum oxide). Small amounts of aluminum oxide are found in slightly impure gem forms such as sapphire and ruby.

Consider the nature of this thin protective oxide film on an aluminum surface. Picture in your mind the surface likened to a landscape covered with clumps of tangled bushes in a forest of trees packed so closely together that no rain reaches the ground. Then, to get a working model of the surface of an aluminum alloy, miniaturize this picture until the height of the forest is a few thousand millionths of an inch tall and consists of clumps and tall spikes of tightly packed crystals of the various metallic oxides. Little wonder that penetration of oxygen and moisture to the surface is difficult, especially after sealing the surface in boiling water and organic dyes to color it.

Chromium and cadmium are reactive metals like aluminum. Each depends on a thin, tightly adherent oxide film for protection from further corrosion. The protective oxide formed on chromium, Cr_2O_3, is rugged, self-healing, and adheres so tightly that corrosion is seldom a problem. The aluminum oxide formed on aluminum is nearly as good, but in the case of iron, the oxide coating formed (on iron) is not as durable.

For a model of an iron surface, visualize a coating of rust as consisting of layers of crystalline iron oxides with fractures and crevices extending to the iron surface. There can be three layers on top of the iron surface. Next to the iron surface there is chiefly ferrous oxide (FeO), grading into the second layer of magnetite (Fe_3O_4), with the top layer of ferric oxide (Fe_2O_3)—the oxide with the most oxygen content because it is next to the air.

5.18 The Corrosion Process

There are many types of corrosion. Each type involves the transfer of valence electrons from one substance to another. The metallic substance that is corroded provides valence electrons which result in another reaction. One reaction cannot occur without the other taking place. Thus in the net galvanic corrosion reaction, two electrode reactions must occur with no electrons unaccounted for. The common types of corrosion are known as galvanic (or uniform corrosion), local (or nonuniform corrosion such as pitting, crevice, and intergranular corrosion), and stress (mechanical corrosion). These differ mainly in the method and the location of occurrence.

5.19 Example of the Oxidation of Aluminum

The spontaneous reaction of aluminum with oxygen in the air can be shown by removing the protective oxide layer from the surface of a sheet of aluminum foil chemically with a drop of a saturated solution of mercuric chloride. A mercury amalgam is then formed to which the aluminum oxide coating is unable to cling. The reaction of the unprotected aluminum with oxygen from the air is exothermic (evolves heat) and spontaneous. Note that some grades of aluminum foil are treated and sealed so tightly that penetration of mercuric chloride to the aluminum surface is difficult, even when scratched.

5.20 Galvanic Corrosion between Dissimilar Metals

Galvanic corrosion occurs when two metals, some distance apart in the potential series, are in electrical contact and are connected by a moist film containing ions forming an ionic path. In effect, the two metals form a battery with their terminals connected, or short-circuited. Current will continue to flow as long as any of the anode (more reactive metal) remains. Unlike the storage battery, these processes are irreversible.

Galvanic corrosion causes a flow of valence electrons, called direct current, to go from the more reactive metal to the other metal. For this current to flow, there must be a return path for the electrons to get back to the more reactive electrode of the battery (cell). This path usually consists of ions of inorganic salts present on the surface of the metals as a film or dirty surface. This is equivalent to connecting the ends of two metal strips with a wire and dipping the other ends into salt water. Electrons can flow through metal, but ions must take their place in water. Positive ions move in one direction; negative ions move in the other direction. At each electrode the ion must give up its charge to the electrode by a chemical reaction.

5.20.1 Harnessing corrosion in a battery

Spontaneous chemical reactions, with valence electrons flowing from one element to another, can be harnessed in a battery, also known as a voltaic cell or galvanic cell. The common dry cells used in flashlights are not dry inside, but moist. Dry cells and lead storage batteries used in automobiles are good examples of chemical batteries that produce a current of electrons that flow only in one direction, called direct current (dc). This is in contrast to the current available at outlets in our homes, called alternating current (ac). Alternating currents cause electrons to reverse direction back and forth (pulsate) 60 times per second (60 hertz) through household appliances.

If the two elements that comprise a galvanic cell are separated with each in its own container, two half-cells are formed. A battery is more easily understood if it is separated into two half-cells. For example, consider the zinc part of a battery in which zinc metal gives up electrons to form zinc ions. These electrons from the metallic zinc in one part (a half-cell) can combine with copper ions to form copper metal in another part of the battery (the other half-cell). This is a spontaneous chemical reaction which converts chemical energy into electric energy. This electric energy can be converted into electromagnetic radiation (light) in a flashlight bulb.

We can physically separate this chemical process into two parts. In one half-cell zinc gives up electrons to form zinc ions. In the other half-cell these electrons change copper ions into copper metal. In practice, the zinc half-cell consists of a strip of zinc metal bent over the rim of a beaker so that it dips into a 1 M (1 molar) solution of zinc sulfate. (A 1 M solution is defined as 1 molecular weight of a chemical compound in grams dissolved in water to make 1 liter of solution.) The other half-cell beaker has a strip of copper metal bent over its rim so that the metal dips into a 1 M solution of copper sulfate.

The electric path between the two beakers is completed by a strip of paper that has been soaked in a salt solution, such as potassium chloride, and hung over the rims of the two beakers, with each end of the paper dipping into one of the solutions. The voltage between the zinc and copper strips of this battery, made of the two half-cells, can now be measured with a sensitive electronic voltmeter. Why is the zinc electrode the negative electrode? Electrons accumulate on the zinc electrode, giving it a negative charge as the zinc spontaneously liberates valence electrons to form zinc ions.

Note that the corrosion process has three essential steps. (1) The electric current flows through the metal electrodes as electrons. (2) The current is carried through the solutions in the beakers and through the filter paper as electric charges on ions. (3) At each electrode, transfer of the charge from ion to metal involves a chemical reaction. At the zinc electrode a zinc atom is converted to a zinc ion and two electrons. At the copper electrode a copper ion and two electrons form a copper atom.

These copper-zinc electrodes, used to illustrate a battery, can also be used to illustrate a sacrificial anode.

5.20.2 Example of galvanic corrosion

A corrosion cell can be made by replacing the electronic voltmeter used in Sec. 5.20.1 with a copper wire with alligator clips on each end. The voltmeter measures the voltage without allowing an appreciable amount of current (electrons) to flow. The copper wire allows the electrons to flow easily and chemical reactions to occur, unchecked, at the electrodes. The zinc electrode releases electrons as the metal reacts chemically (corrodes) to form zinc ions. The current is carried through the solution by the positive zinc and copper ions and by the negative sulfate ions. The net result is corrosion of the zinc electrode with copper metal being deposited on the copper electrode. In the absence of copper ion, hydrogen ion from water is converted to hydrogen gas at the less-reactive electrode. In the corrosion of iron to form rust, hydrogen gas is formed on the surface of the other less-reactive metal electrode touching the iron.

5.20.3 Example of a sacrificial anode

An aluminum half-cell can be made by filling a small beaker two-thirds full with a 1 M aluminum chloride ($AlCl_3$) solution and bending a strip of aluminum over the rim of the beaker so that the strip reaches into the solution to the bottom of the beaker. This will be the sacrificial anode. Then by combining this half-cell with another half-cell made in the beaker with an iron strip dipping into a 1 M solution of table salt (NaCl), a realistic sacrificial anode is produced. The electric circuit must be completed with a copper wire connecting the iron and aluminum electrodes. In this cell the aluminum will be converted to aluminum ions and electrons. These electrons left behind on the electrode will give it a negative charge, making it the anode.

Note that the aluminum is converted (corroded) to form aluminum ions instead of the iron electrode corroding. Thus the aluminum is corroded (sacrificed) to preserve the iron. Hydrogen gas is formed at the iron electrode.

5.21 Standard Electrode Potential Series

The voltage obtained from spontaneous reactions involving the flow of electrons from one element to another in a galvanic cell is a measure of the spontaneity of the chemical reaction, such as corrosion. Much can be learned by separating the reactants of these cells into half-cells. The electronic potential energy of these half-cells can then be measured in volts and the data put into a table.

A zinc half-cell would consist of a strip of zinc in a solution of zinc ions in a separate container. Notice that even a gas, such as hydrogen, when bubbled through a solution, can give its valence electrons to an inert platinum strip to form a hydrogen electrode half-cell.

Because it is impossible to measure the voltage of a half-cell (or half of a battery), it is customary to measure the voltage of a half-cell coupled to a standard electrode. Table values are usually given with the standard hydrogen electrode as the other half-cell. By assuming that the hydrogen electrode has zero voltage, the entire voltage of the cell is assigned to the other electrode as its half-cell voltage.

TABLE 5.2 Standard Electrode Voltages

Electrode reduction reaction	Voltage, V
$Li^+ + e = Li$	−3.05
$K^+ + e = K$	−2.93
$Na^+ + e = Na$	−2.71
$Mg^{++} + 2e = Mg$	−2.37
$Al^{+++} + 3e = Al$	−1.66
$O_2 + 2H_2O + 4e = 4OH^-$	−0.83
$Zn^{++} + 2e = Zn$	−0.76
$Cr^{+++} + 3e = Cr$	−0.74
$Fe^{++} + 2e = Fe$	−0.44
$Cd^{++} + 2e = Cd$	−0.40
$Ni^{++} + 2e = Ni$	−0.28
$Sn^{++} + 2e = Sn$	−0.14
$Pb^{++} + 2e = Pb$	−0.13
$Fe^{+++} + 3e = Fe$	−0.02
$2H^+ + 2e = H_2$	0.00
$Cu^{++} + 2e = Cu$	0.34
$Ag^+ + e = Ag$	0.44
$Hg^{++} + 2e = Hg$	0.80
$Cl_2 + 2e = 2Cl^-$	1.36
$Au^{+++} + 3e = Au$	1.50

The voltages of electrodes relative to the hydrogen electrode are shown in Table 5.2. Note that the more common term of "voltage" is often used in place of "potential." The voltages in this table are standard voltages. This means that the metal electrode is immersed in a standard 1 M solution of its ions. (A standard 1 M solution is 1 mole of the substance dissolved in 1 liter of water.) Gases are at a standard pressure of 1 atmosphere. The voltages are measured at equilibrium conditions. Half-cell equations with electrons on the left side, or the reactant side of the equation, are called reduction reactions. An oxidation reaction is indicated if the reaction is turned around so that electrons are products. The equations are written in skeleton ionic form to show only the substances that are chemically changed in the reaction.

The standard electrode potential series in Table 5.2 can be used as a chemical displacement series. Any metal on the right side of the equation will displace, or replace, an ion in any equation below it by converting the ion to the element. Lithium, potassium, and sodium are exceptions because they liberate hydrogen from the water that is present. The table also serves as an activity series with the most chemically active elements at the top of the table and the most unreactive elements at the bottom. Going down the table, the elements decrease in their tendency to exist as ions.

The tendency of an ion to take back its valence electrons and return to the element increases as you move from the top to the bottom of the table, as shown by the change in the voltage from negative to positive. Gold, at the bottom of the table, is so stable that it occurs as the complete metal in nature. The elements at the top of the table will liberate hydrogen gas from hydrogen ions, but with less vigor, until hydrogen itself is reached.

The equations represent half of a reaction since they list electrons as a chemical ingredient. Any two of these equations can be added to give a balanced chemical equation. Turn one of them around before adding them. However, all of the electrons lost by one element must be gained by the other one. To get the same number of electrons in each half-cell, either one of the equations may be doubled or tripled. In this respect, chemical equations are treated like algebraic equations.

5.22 Local or Nonuniform Corrosion

Metals have dense, closely packed crystal structures. Crystal imperfections cause them to group in clumps or grains, which can be seen under a microscope by etching away the more reactive boundaries with acid. Stress hardening (cold work) involving excessive stress or flexure produces crystal distortions and fractures as metal atoms slide over one another. Ferrous sulfide (FeS) and cementite (Fe_3C) have low melting points and solidify around these grains to form corrosion sites.

Intergranular corrosion results when one crystal face becomes an anode and another the cathode. Other causes for nonuniform corrosion in steel can be identified. Otherwise, the location of these anodes and cathodes is a matter of chance. A small amount of surface dirt can be responsible for starting concentration-cell anodes. The electric circuit between these tiny anodes and cathodes, naturally provided by the metal, must be completed by a surface film of ionic conductors. Prevention of this common type of corrosion is easier than its correction once it starts.

Local nonuniform corrosion is more varied and more difficult to prevent than uniform galvanic corrosion. Often caused by slight differences in surface conditions, localized corrosion starts from a multitude of microscopic anodes. Pitting or crevice corrosion can start when a bit of soil causes a change in surface ion concentration, when there is moisture, or when there is access to oxygen from the air. In addition to this type of concentration cell, intergranular and stress corrosion is common.

Local corrosion can also be caused by mechanical stress or cracks in the metal surface. The stress energy causes the areas of high stress on a metallic surface to corrode and be more reactive (anodic) compared to the less-strained areas. Another common form of local corrosion occurs in a scratch or crack in the metallic surface. The crevice collects dirt and moisture, which produce a concentration cell and cause local corrosion in the bottom of the crevice.

In atmospheric corrosion oxygen and moisture play an essential role. Moisture is a regulating factor in the corrosion of metal surfaces. Figures 5.11 and 5.12 show structures subject to atmospheric corrosion. A similar condition occurs if the metal surface is totally immersed in water. Oxygen can only be replenished at the submerged surface by slow diffusion through water. This explains the rapid corrosion that occurs on metals at the water-air interface. This type of corrosion occurs when metal structures are anchored in the ground. Corrosion is rarely a problem in dry environments when the relative humidity of the air is low.

Figure 5.11 Controversial architecture of the Pompidou Center, Paris, France, 1985, showing the external major supports and air vents.

Figure 5.12 The Eiffel Tower with its network of steel beams and an elevator is visible from any part of Paris. Built in 1889, this marvel of construction presents a constant challenge to corrosion engineers.

5.23 Corrosion-Resistant Metals and Alloy Steels

By making an alloy of aluminum with small amounts of copper, both the tensile strength and the corrosion resistance are increased.

Stainless-steel alloys containing chromium have good corrosion resistance if they are properly protected. If the expense of Monel, stainless, and other specialty steels is impractical, chemical-corrosion inhibition is needed, such as a protective organic or inorganic coating.

Corrosion can occur on the surface of sections of a corrosion-resistant metal or alloy. In many environments metals are exposed to a corrosive atmosphere of moisture, salt particles, acid rain, other air pollutants, and, of course, oxygen. In addition to small irregularities in the metal surface, inaccessible areas exist that collect pockets of dust. Each of these is a potential anode or cathode for pitting in corrosive environments, especially when the relative humidity is 60 percent or more. Once started, pitting is autocatalytic and feeds on its own corrosion products.

Oxygen, available for the pitting reaction, decreases as the pit forms due to slow penetration from the surrounding atmosphere. On the other hand, chemical salts from corrosion are protected from the weather and can exceed the 1 M concentrations listed in the galvanic series table.

For example, the concentration of ions found in the bottom of pits, in crevices on the surface, or in cracks in the iron oxide crust can vary enormously from the neutral pH of 7. The pH of the corrosive chemical products found at the bottom of these anodic corrosion pits can be as acid as 0.001 M H^+, or a pH of 3, due to the reaction of the metallic ions such as Fe^{++} with water, as shown by the hydrolysis equation

$$Fe^{++} + 2H_2O \rightarrow Fe(OH)_2 + 2H^+$$

Each tenfold change in concentration of one of the ions in the potential series equation changes the voltage by 0.06 volt divided by the number of electrons in the electrode equation. For the hydrogen electrode, a tenfold drop in H^+ concentration from the standard 1 M value, or 1 pH unit, changes the voltage to -0.06 volt. For a pH of 3, the voltage becomes -0.18 volt. The acid produced diffuses to the top of the pit, so pitting can be detected by pressing moist indicator paper to the metal surface.

Stress cracking, such as intergranular attack and pitting, is a dangerous form of localized corrosion. Failure may come without warning. The combination of stress and corrosion is synergistic, in which the effect of the combination is greater than the sum of the two. Stress cracking can result from localized corrosion reducing the tensile strength of the metal by as much as 90 percent of its normal value.

Extensive corrosion occurs on metal supports at the line of contact with moist soil. This is due to the oxygen concentration cell that is formed. Since the soil contains less oxygen than the air above it, the surface of the metal just below the surface becomes the cathode with the evolution of hydrogen, whereas the line at the surface becomes the anode and corrodes to form the metallic ion. This type of concentration-cell corrosion also occurs with pipes buried under sidewalks or other covering. Again, the metal exposed to oxygen in the air is the cathode with the more inaccessible pipe corroding as the anode in the oxygen-deficient soil. Note that if more than one electrode reaction is possible, the one with the largest positive net voltage for the existing ion concentrations will occur.

5.24 Don't Give Up

Lest you get the feeling that nature is hungrily waiting to corrode and destroy all that we produce and we are powerless to interfere, don't give up hope. It is necessary to understand the problem before arriving at a solution. Section 5.25 tells how to meet the challenges.

5.25 Balancing Redox Equations

The term "redox" is an abbreviation for an oxidation-reduction process. Oxidation-reduction is the transfer of valence electrons from one element to another, that is, one element is oxidized (gives up valence electrons) as another element is reduced (takes up these lost electrons). Note that oxidation of an element cannot take place without another element being reduced concurrently. Thus oxygen is not required, in spite of the use of the term "oxidation."

The half-cell reactions in Table 5.2 can be used to write balanced chemical equations and to predict whether the reactions are spontaneous. For example, what will happen, if anything, if you put a strip of aluminum metal in a 1 M solution of ferrous iron ions? First, write the reaction for the ferrous iron half-cell as given in Table 5.2. Next, write below it the reverse of the equation for the aluminum half-cell. Reverse the sign of the voltage from − to +. Add the two equations as shown:

$$Fe^{++}(aq) + 2e = Fe(s) \qquad -0.44 \text{ volt}$$

$$Al(s) = Al^{+++}(aq) + 3e \qquad +1.66 \text{ volts}$$

$$Al(s) + Fe^{++}(aq) = Al^{+++}(aq) + Fe(s) + e \qquad +1.22 \text{ volts}$$

The symbol s refers to the solid; aq refers to the ion in water (aqueous) solution. The symbol e, appearing in the net equation, stands for a valence electron—it is neither an element nor a compound, so it does not belong in a balanced chemical equation. What should be done so that each half-cell has the same number of electrons to be canceled? The answer follows.

If both sides of a half-cell equation are multiplied by the same number, they will still be equal. Each equation will have six electrons if the first one is multiplied by 3 and the second one by 2. The voltage of the equation is unchanged because the voltage of a battery is not changed by increasing its size. By applying these factors, the six electrons can be canceled:

$$3\{Fe^{++} + 2e = Fe(s)\} \quad -0.44 \text{ volt}$$

$$2\{Al(s) = Al^{+++} + 3e\} \quad +1.66 \text{ volts}$$

$$3Fe^{++} + 2Al(s) = 3Fe(s) + 2Al^{+++} \quad +1.22 \text{ volts}$$

Since the net voltage is positive, the reaction of aluminum with a 1 M solution of ferrous iron will be spontaneous from left to right.

5.26 Displacement Series

The standard electrode voltage series can be used as a displacement or replacement series. Any element in the series will displace or force any ion below it to return to the element. That is, a strip of copper metal placed in a solution of silver ions will be coated with silver metal as copper ions form. Another example is the prediction that aluminum metal, in a solution of ferrous ions, would be coated with iron metal as aluminum ions are formed. Figuratively speaking, any reaction in Table 5.2 takes place, spontaneously, as it reverses a reaction above it. This is shown by the arrows drawn in the following example:

$$Cu^{++} + 2e = Cu \leftarrow$$

$$Ag^+ + e = Ag \rightarrow$$

5.27 Prediction of Galvanic Corrosion

Using Table 5.2, a balanced chemical equation can be written for the reaction of iron with a zinc sacrificial electrode. Theoretically, any metal could be protected from galvanic corrosion if connected to a metal above it in the table. Writing the reversed half-cell equation for the oxidation of zinc from the table above the equation for the hydrogen half-cell gives the following net balanced equation:

$$Zn = Zn^{++} + 2e \quad 0.76 \text{ volt}$$

$$2H^+ + 2e = H_2 \quad 0.00 \text{ volt}$$

$$Zn + 2H^+ = Zn^{++} + H_2 \quad 0.76 \text{ volt}$$

Notice that if the reduction equation for sodium, instead of that for hydrogen, had been used, the net result would have been the same because sodium metal would have immediately reacted with water to produce hydrogen.

Since a positive voltage for a balanced equation means the reaction is spontaneous from left to right, a negative voltage means that the reverse is true and the equation should be turned around to be spontaneous from left to right. Thus for the balanced equation,

$$Fe + Cu^{++} = Fe^{++} + Cu \quad + 0.78 \text{ volt}$$

The positive voltage shows that the reaction of iron in 1 M cupric ions is spontaneous to form ferrous ions and copper. It also shows that copper can be protected from corrosion by connecting it to scrap iron with an electric wire. Iron is a sacrificial anode in this application.

Consider the result if zinc and iron, in mechanical contact, are immersed in 1 M solutions of their ions. Copy the half-cell equation for zinc from Table 5.2. The equation for the iron electrode is turned around so that the electrons can be canceled. When the equation is turned around, the sign of the voltage must be reversed. The net equation is

$$Zn^{++} + Fe = Zn + Fe^{++} \quad -0.32 \text{ volt}$$

The negative sign for the voltage indicates that the reaction should take place, spontaneously, from right to left. To be spontaneous from left to right, the reaction should be written:

$$Zn + Fe^{++} = Zn^{++} + Fe \quad +0.32 \text{ volt}$$

Fortunately, steel can be doubly protected from corrosion by a galvanizing process in which it is dipped in molten zinc. Zinc forms a better protective oxide coating than iron and is more corrosion-resistant. Zinc also serves as a sacrificial anode when it is eaten through and the iron surface is exposed.

5.27.1 Example of a concentration cell

Make a duplicate of the copper electrode half-cell used in Sec. 5.20.1. For comparison with the concentration cell, measure the voltage between the two copper electrodes with the electronic voltmeter. Any measurable voltage between these identical cells will be small and due to the difference in the mechanical treatment of the copper electrodes.

Prepare a 0.1 M copper sulfate solution by mixing 20 milliliters of 1.0 M with 180 milliliters of distilled water. Replace the 1.0 M solution in one of the beakers with this 0.1 M copper sulfate solution. The voltage of this cell should be near the theoretical value of 0.26 volt.

Note: The second law of thermodynamics operates to bring the two copper solutions toward the same concentration by reducing copper ion to copper metal on the electrode in the more concentrated solution; the reverse reaction occurs at the anodic electrode. The net reaction is the transfer of copper ions from the more concentrated solution to the more dilute one.

5.27.2 Sacrificial anode cell

Replace the copper electrode half-cell of the previous galvanic-cell exercise with a beaker containing an iron electrode in a 1 M solution of table salt. To demonstrate the corrosion of iron in a salt solution, immerse the end of another strip of iron in the same beaker (as a control for comparison). Notice that both strips of iron will slowly start to rust. Connect one of the iron strips to the zinc strip in the other half-cell with alligator clips on each end of a copper wire. The iron electrode is protected and the zinc electrode is sacrificed as zinc ions and electrons are formed. Note that the oxidation of the zinc electrode makes it the sacrificial anode. Hydrogen is evolved at the negative iron electrode.

5.28 Surface-Corrosion Protection

Proper treatment of metals has much to do with their resistance to corrosion. Surfaces are passive, or inactive, if the ionic flow between anode and cathode is restricted by a surface barrier on one of the metals. Polarization usually refers to a depletion of ions, an oxide coating, or a gas layer on the metal surface. This decrease in the corrosion current can be produced by using inhibitors that aid the natural process or by prior surface treatment.

Some metals, such as tin, have a natural tendency to polarize. Chromium also acts as an inhibitor that promotes polarization by forming a durable film as it becomes the anode. This protective oxide film can be augmented, if necessary, in a low-oxygen environment by treating with an oxidizing agent such as a chromate. Again, it should be noted that stainless steel and the more corrosion-resistant Monel and Inconel are vulnerable to pitting and crevice corrosion in the high chloride concentrations found in ocean environments.

5.29 Corrosion Protection from Circulating Water

Water has an unusually large heat capacity of 1.0 calorie per gram per degree Celsius. (One calorie of heat is required to raise the temperature of 1 gram of water by 1°C.) When used in automobile radiators, water contains oxygen from the air in addition to dissolved salts. Sodium chloride and oxygen are active corrosion agents, and it is usually impractical to remove one of them to prevent corrosion. Corrosion retardants contain oil, detergents, and sodium chromate. Although ethylene glycol has only about half the heat capacity and only half the cooling capacity of water, it is combined with water for use in radiators because of its lower volatility and antifreeze protection.

For closed water systems in space heating and industrial cooling, anticorrosive treatment consists of softening the water by chemical means or use of deionized water. However, even pure water causes corrosion of many metals due to its oxygen content. Pitting is thought to occur because of partial shielding of the surface from oxygen by particles of debris. Traces of oxygen can be removed by the addition of sodium sulfite, which reacts with oxygen to form sodium sulfate. This is a slow reaction and can be speeded up by the addition of a small amount of a copper (or cobalt) salt which acts as a catalyst. (A catalyst temporarily enters into a reaction, but is restored and can be reused.) If the water system to be protected is extensive, additional protection can be provided by electrochemical means or by using equipment with a corrosion-resistant ceramic lining.

5.30 Metal Cladding

Although cast iron, stainless steel, Monel, and Inconel are corrosion-resistant, they have limitations and are expensive. One alternative is to give steel a metallic protective surface, called metal cladding.

A coating of copper, chromium, or nickel on steel provides a cathodic surface that is corrosion-resistant. Or a coating of aluminum, tin, or zinc can provide anodic protection. Thin sheets of these metals can be coated onto steel by cladding. This is done by hot dipping the steel into one of these molten metals, by hot rolling a sheet of one of the metals onto the surface, or by electroplating. To protect the molten metal from reacting with the oxygen in the air in the dipping process,

Figure 5.13 Galvanized steel framing with bracing, roof trusses, and rafters in place.

the surface of the molten metal must be covered with a flux. Molten zinc applied to steel forms a surface alloy that helps bond the zinc to the surface. Steel clad in this way with zinc is called galvanized. An example of galvanized steel framing is seen in Fig. 5.13.

5.31 Electrodeposition of a Metallic Coating

Most common metals can be electroplated. (Aluminum is an exception.) The result is a thin, more uniform coating with fewer holes or other imperfections that occur with dipping. Metals listed above iron in the displacement series in Table 5.2 give additional corrosion protection, even if the coating is broken. Note that aluminum metal is prepared by the electrolysis of a molten mixture of aluminum salts.

Electroplating requires the use of additives in the plating solution to provide a uniform, smooth adherent coating. For example, copperplating baths contain a high concentration of either sulfuric acid or sodium cyanide. Cyanide reduces the copper ion to the optimum concentration for plating by forming a copper cyanide complex ion. This ion dissociates in order to liberate more copper ions to replace those electroplated from the bath. To obtain adherence, surface preparation is needed to remove oil and rust.

5.32 Galvanic Protection with Sacrificial Anodes

To protect metals from electrochemical corrosion, the contact of dissimilar metals should be avoided. This can be seen by using Table 5.2 to find the voltage differences between metals responsible for corrosion. The larger the difference in voltages, the more vigorous the corrosion process. If close proximity is unavoidable, there are two alternatives.

The first alternative is electrical insulation between dissimilar metals. Insulation can be used in the design to reduce the flow of electrons between anodic and cathodic metals. Bolts holding dissimilar metals together should be separated with insulating sleeves and washers. To avoid a feeling of false protection, care must be taken in both the planning and the installation not to create crevices for the accumulation of dirt, with its ionic paths around the insulators.

The second alternative is expensive and is used only if the structural metal or pipeline is in contact with the ground or is buried in the ground. In these cases, galvanic protection should be provided with either a sacrificial anode or an applied voltage. A metal higher in the electrode potential table makes a suitable anode, such as aluminum, zinc, or magnesium.

In practice both the shape of the sacrificial anode and its distance from the metal to be protected are important. A metal buried in moist ground dissolves faster around the edges, or if it is in an area that is under stress, or at the point where the electric wire is attached to the sacrificial anode. The part of the pipeline nearest the sacrificial anode will receive the most protection. The anodes should not be more than 50 to 100 feet away from the pipeline because of the high resistance of the electrical path provided by the ground between the pipeline and the anode.

5.33 Example of a Zinc Sacrificial Anode

Part A. Corrosion between dissimilar metals

Preparation. Bend a strip of iron over the rim of a 250-milliliter beaker. Bend a strip of copper metal over the rim of another beaker. Prepare a test solution by dissolving 50 grams of sodium chloride, 1 gram of potassium ferricyanide, and 1 milliliter of phenolphthalein solution in 600 milliliters of water. Pour 200 milliliters of the test solution into each beaker.

Directions. Connect a high-resistance milliammeter, or an electronic multimeter with alligator clip leads, to the metal strip electrodes. Connect the negative lead of the meter to the iron (negatively charged) strip. No metallic corrosion will occur until the electric circuit is completed. Dip a strip of filter paper into the test solution and hang it over the rims of the two beakers so that it reaches into the solution in each beaker. Note that a small current now flows through the milliammeter. Observe that the test solution turns blue near the iron electrode, showing the presence of ferrous ions produced by the anodic dissolution of iron. This shows that galvanic corrosion of the iron strip has taken place. If the high resistance of the meter prevents enough electrons to flow to produce a color at the anode, replace the meter with a copper wire. Notice that you have made a battery. In time, a visible etching will occur on the iron strip.

Discussion. The salt solution in the filter paper completes the electric circuit. Electrons flow from the iron strip through the meter to the copper strip, but electrons can return to the iron strip only through the solution as electric charges on the ions. This completes the electric circuit. Since electrons at low voltage cannot flow through air or water, they must pass through the solution either as negatively charged ions or by the equivalent process of positive ions going in the opposite direction.

If the current is allowed to flow long enough, a pink color will be visible around the copper cathode, showing that, in the absence of copper ions, hydrogen ions from water are converted to hydrogen gas, thus freeing the hydroxide ions from the water.

Part B. Corrosion protection with a zinc sacrificial anode

Directions. Bend a strip of zinc over the rim of another 250-milliliter beaker containing the sodium chloride test solution left over from part A. Use the iron strip electrode half-cell for the other half of the corrosion cell. As before, connect the milliammeter to the metal electrodes with the positive lead of the meter connected to the iron strip. Complete the circuit with a strip of paper dipped into the test solution.

Observe that the iron strip no longer has anodic corrosion. Blue ferro ferricyanide does not form, but after a short while a precipitate of zinc hydroxide will be observed around the zinc strip. This shows that the zinc anode is being sacrificed to avoid corrosion of the iron. Since the sodium chloride solution has a neutral pH of 7, zinc hydroxide is insoluble.

Interpretation. The zinc metal provides electrons that are conducted through the meter to the iron electrode. The negatively charged iron surface converts positively charged hydrogen ions from water to hydrogen gas, thereby freeing hydroxide ions from the water. Notice that for electrons to complete their circuit, a chemical reaction must occur at each electrode.

5.34 Sacrificial Metallic Pigment Coating

Aluminum and zinc are often used in organic coatings for anodic protection. Other metallic oxidizing pigments used in primer undercoats as corrosion inhibitors are zinc chromate, red lead, and litharge. Zinc chromate is used instead of the more water-soluble sodium chromate that is more easily removed from the organic coating by rain or water immersion. If the chromate is too insoluble, an insufficient amount of chromate ion will be available in the coating to provide the needed protective oxide-film passivation.

5.35 Corrosion Protection with Applied Voltage

It is possible to replace the sacrificial anode with a direct-current electric generator. A few volts, equal to the voltage provided by a sacrificial anode, should be all that is necessary. However, due to the high current needed to passivate pipelines, tanks, and other structures buried in the ground, a much larger voltage is required. Any metal can be made the positive anode by applying a high enough voltage to reverse any natural tendency. Then less expensive materials, for example, scrap steel, Duriron, lead, graphite, or even a noble metal such as platinum, can be used.

The negative lead from the generator, or rectifier, is connected to the metal to be protected, making it the cathode. Due to the lack of more easily electrolyzed ions, water is converted to hydrogen gas and hydroxide ion at the protected metal cathode. The hydrogen formed can dissolve in the surface of some metals to produce a problem called hydrogen embrittlement by reducing the tensile strength of the metal being protected. It would not help to coat the metal with an organic coating because it would be popped off by the hydrogen gas.

As in the sacrificial anode method, the placement of the many buried anodes is important in order to provide a uniform cathodic voltage over the entire surface of the protected metal. The cathodic corrosion protection method requires experience in design and maintenance in order to ensure efficient and complete protection and to avoid raising the voltage unnecessarily high so that it becomes a hazard to small children and animals straying into the area.

5.36 Passivation by Anodizing

The normal surface oxide layer that acts as a barrier to corrosion can be increased in thickness up to 100 times. To further passivate a surface, an anodizing process is used that increases the thickness of the oxide coating. In the usual anodizing process, aluminum is made the anode by attaching it to the positive pole of a battery, or rectifier, in a hot dilute solution of sulfuric acid. The limiting thickness of the oxide layer that can be artificially grown in the anodizing process depends on the temperature and concentration of the solution. A mechanical rupture of the film may occur if it is made too thick.

In this process it is thought that the negatively charged oxide ion moves through the oxide coating in addition to the movement of the positive aluminum ion from the aluminum surface. The two ions meet in the coating to form aluminum oxide, replicating the atomic arrangement in the substrate. Stresses formed in the anodized coating are small because the oxide formed occupies a slightly larger volume than the metal. The film is stable and will not buckle and pull loose from the surface. This tightly adhering film is needed for a protective anodized coating.

A porous oxide coating is desirable for a decorative coating so that the pores can be filled with colored pigments or dyes. Any porous structure remaining is then sealed by steam or boiling water. It is also possible to select an aluminum alloy in which the additives form the colored compound on anodizing. Anodized film on aluminum is tough, coherent, and corrosion-resistant.

5.37 Organic Protective Prime Coat

Prime coats on metals are used to promote adhesion of the top coat as well as to provide corrosion resistance. Adhesion can be improved by a slight chemical etching of the metal and by the use of a prime coat containing phosphate or chromate, which alter the surface chemically. Silicates and phosphates are used to decrease the porosity of the oxide coating on aluminum.

Iron or steel is not benefited by chromate treatment as much as aluminum because the oxide formed on the surface is not as tightly adherent as on aluminum. However, use of a wash primer on aluminum and iron (containing dilute phosphoric acid) does form a tenacious phosphate surface film that improves the adhesion of an organic coating.

5.38 Phosphoric Acid Treatment of Metallic Surfaces

Phosphates are reported to have been used by the Romans to protect iron as early as the third century. In 1910, iron and steel were treated with a phosphoric acid and ferrous sulfate solution. The Parker Company improved this treatment in 1918 by replacing the ferrous sulfate with manganese phosphate. In this process, known as parkerizing, manganese dioxide was added to oxidize some of the iron to ferric phosphate. The time required for this treatment was further reduced (from 3 or 4 hours for the parkerizing treatment) to 10 minutes by the use of manganese dihydrogen phosphate plus a small amount of copper salt in the formula. This improved process, now known as bonderizing, requires less than 1 minute and can be used on steel, zinc, and aluminum surfaces.

The phosphating process involves dipping iron or steel into a bath containing a nearly saturated water solution of zinc dihydrogen phosphate in a phosphoric acid solution. This is said to produce zinc phosphate eventually on the metal surface by the following reaction, where M is bivalent zinc or iron,

$$3M(H_2P_4)_2 = M_3(PO_4)_2 + 4H_3PO_4$$

An insoluble layer of fine crystals of iron or manganese phosphate increases the adhesion of paint to the surface and adds some corrosion protection. To speed the action of these phosphate salts the following accelerators are used: copper nitrate, copper nitrite, copper chlorate, nitrobenzene, and toluidine.

5.39 Corrosion-Protective Pigments

Iron oxide has been used as a pigment for many centuries and is now widely used in prime coats on ferrous metals. Of the pigments, ferric oxide is red, hydrated ferric oxide is yellow, and hematite (Fe_3O_4) is black. Iron oxide is used in the prime coat because it is opaque to both the visible and ultraviolet lights and reduces the permeability of the coating to moisture and corrosive salts. However, neither iron oxide nor rust have been found to promote or hinder corrosion.

Metals in the subdivided state, such as aluminum and zinc, can be used in paint to provide corrosion-protective pigments. These are especially effective if they are in a flaky or leaflet shape that can overlap to form a barrier to moisture and corrosive salt migration.

A wash primer was developed for use during World War II, patented in 1950, and has been popular for many years for the protection of steel and aluminum. This WP1 wash primer was designed to combine some of the best technology at the time in a two-package preparation. Part A contains poly(vinyl acetate) in alcohol as a film former with a zinc phosphate pigment. Part B consists of water, alcohol, and phosphoric acid. This can be classified as a type of conversion coating since it reacts with the metal substrate to reduce corrosion. It mildly etches the surface and reacts to form a phosphate, both of which increase adhesion.

The addition of zinc chromate provides some corrosion protection to a primer as it converts the film former to a more durable coating until the top coat can be applied.

Lead pigments have the embarrassing property of reacting with traces of hydrogen sulfide (from air pollution) to form black lead sulfide, especially in the vicinity of oil wells and thermal power sites. Because of the environmental harm from lead and chromium compounds, these are being replaced by a two-package epoxy paint that forms a good barrier to water and corrosive ions, but with a short pot life before use.

Red iron oxide is sometimes used as a red pigment in prime coats but offers little if any corrosion protection. Corrosion inhibitors that have been shown to be effective on iron or steel are red lead and chromates such as zinc chromate.

Although still used in industry, in the U.S. Navy primers are being replaced by epoxy coatings, which have the advantage of providing a superior barrier to water and corrosive salts without containing chemicals toxic to the marine environment. Epoxy paints have unusually good adhesion and barrier properties, but are expensive and require trained painters because of the hazardous nature of the ingredients.

5.40 Chemical Conversion Coatings

Primers that react with the surface metal to form a protective surface coating are called conversion coatings. Chromates, phosphates, and fluorides are commonly used to protect aluminum, magnesium, and steel. These coatings not only improve the corrosion resistance of the metal surface, they also give the surface a slight etch to improve the adhesion of the top coating of paint.

5.41 Building Problems Related to Metals

Metallic corrosion can be kept to a minimum by using certain precautions. Cavities, crevices, surface irregularities, and contact between dissimilar metals are to be avoided in the design of exposed surfaces. Metallic surfaces must be accessible for inspection and treatment. To make full use of corrosion technology, preventive measures should be incorporated at the design stage. This is vitally important if there is exposure to a marine environment, high humidity, or corrosive chemical emissions.

When using corrosion-resistant materials, allowance should be made for the synergistic effects of unusual metal stress, flexure, and fatigue on corrosion. Avoiding the expense of specialty materials by designing to the maximum stress level is a dangerous temptation. Adequate drainage is essential in the design of flat roofs, around metal joints, or in other areas that can collect water and dirt which will cause early corrosion or related failure. Poor welding practices can destroy the corrosion resistance of specialty alloys.

Methods of corrosion protection can be grouped under chemical treatment, electrochemical prevention, and environmental protection. The choice of method must be made on an individual basis, depending on the situations encountered.

Terms Introduced

Allotropic iron
Alloy
Alpha iron
Aluminum alloys
Aluminum protective oxide
Annealing
Austenite
Bainite
Body-centered cubic
Cementite
Copper alloys
Displacement series
Eutectic
Face-centered cubic

Ferrite
Galvanic corrosion
Gamma iron
Half-cell
Hexagonal close-packed
Interstitial solid solution
Iron-carbon phase diagram
Iron alloys
Iron carbide
Iron ore smelting
Lead alloys
Martensite
Metal composites
Metallic bonding

Metallic compounds
Nonferrous alloys
Pearlite
Phase diagram
Quenching
Redox equations

Sacrificial anode
Standard electrode voltages
Steel heat treatment
Substitutional solid solution
Tempering
Work hardening

Questions and Problems

5.1. What is the difference between a solid solution and a molten mixture of two metals?

5.2. Explain the differences in results that occur for slow cooling and rapid cooling of an alloy that has been melted.

5.3. Describe the annealing process of an iron alloy.

5.4. Use a phase diagram to explain the quenching of an alloy.

5.5. What do the numbers stand for in 18-8 stainless steel and why does this steel not rust?

5.6. Describe two methods for corrosion protection of a closed water circulating system.

5.7. Describe galvanizing with its advantages and disadvantages.

5.8. Discuss cladding and the corrosion resistance it imparts to steel.

5.9. Discuss electrodeposition as an alternative to cladding, including the advantages and disadvantages.

5.10. Describe the need for a corrosion-retardant prime coat.

5.11. Explain why a flexible iron wire gets warm to the touch and finally breaks apart when rapidly flexed back and forth at one place. If the wire flexing is stopped soon after the wire has stiffened at the flex point, then heated red hot at the point of flexure, why should the flexed point on the wire regain much of its flexibility?

5.12. Compare the corrosion protection provided by the oxide film on aluminum and on iron.

5.13. Describe the anodizing process and explain the need to follow with a chromate treatment.

5.14. Compare the corrosion protection provided by metal cladding with that obtained by electropassivation.

5.15. Compare the use of a sacrificial anode for corrosion protection with electropassivation.

5.16. Why is it possible to apply too large a voltage for electrochemical corrosion protection?

5.17. Explain the difference between interstitial and substitutional solid solutions.

5.18. Why should a spontaneous phase change be expected to be exothermic?

5.19. Sketch a possible phase diagram for two metals that form a compound in addition to eutectics.

Suggestions for Further Reading

J. T. Atkinson and N. H. VanDroffelaar, *Corrosion and Its Control,* National Association of Corrosion Engineers, Houston, Tex., 1982.

N. A. Bui, F. Irhzo, F. Dabosi, and Y. Limouzin-Maire, "On the Mechanism for Improved Passivation by Additions of Tungsten to Austenitic Stainless Steels," *Corrosion,* no. 12, pp. 491–495, Dec. 1983.

Ulick R. Evans, *An Introduction to Metallic Corrosion,* 3d ed., American Society for Metals, Metals Park, Ohio, 1982.

E. Mattsson, *Basic Corrosion Technology for Scientists and Engineers,* John Wiley & Sons, New York, 1989.

James F. Shackelford, *Introduction to Materials Science for Engineers,* Macmillan, New York, 1985.

C. O. Smith, *The Science of Engineering Materials,* 3d ed., Prentice-Hall, Englewood Cliffs, N.J., 1986.

Peter A. Thornton, *Fundamentals of Engineering Materials,* Prentice-Hall, Englewood Cliffs, N.J., 1985.

Herbert H. Uhlig and R. Winston Revie, *Corrosion and Corrosion Control,* John Wiley & Sons, New York, 1985.

Jan H. L. Van Linden, ed., *Second International Symposium: Recycling of Metals and Engineered Materials,* Minerals, Metals and Materials Society, Warrendale, Pa., 1990.

Lawrence Van Vlack, *Materials for Engineering: Concepts and Applications,* Addison-Wesley, Reading, Mass., 1982.

Chapter 6

Wood, Polymers, Adhesives, and Plastics

6.1 Introduction	174
6.2 Classification of Woods	175
6.3 Composition of Wood	175
6.3.1 Cellulose	175
6.3.2 Pentosan and Lignin	176
6.4 Structural Panel Composites	176
6.4.1 Plywood	176
6.4.2 Particleboard	177
6.5 Bacterial Destruction of Wood	178
6.6 Fungal Destruction of Wood	178
6.6.1 White Rot	179
6.6.2 Brown Rot	179
6.6.3 Soft Rot	179
6.6.4 Rot Protection	179
6.7 Deterioration by Insects	180
6.7.1 Wood Inhabitants	180
6.7.2 Carpenter Ants	181
6.7.3 Carpenter Bees	181
6.7.4 Termites	182
6.8 Prevention and Control of Termites	184
6.8.1 Basaltic Termite Barrier	185
6.8.2 Cell Treatment of Wood	185
6.8.3 Protection by Fumigation	186
6.9 Paper Technology	186
6.9.1 Modern Papermaking	186
6.9.2 Self-Destruction of Modern Paper	187
6.9.3 Treatment of Acid Paper	187
6.10 Polymers	187
6.10.1 Natural and Synthetic Polymers	188
6.10.2 Structure of Linear Polymeric Chains	188
6.11 Physical Properties of Polymers	190
6.11.1 Permeability of Polymers	190
6.11.2 Coefficient of Expansion	192

6.11.3 Glass-Transition Temperature	192
6.11.4 Lowering the Glass-Transition Temperature	192
6.12 Example of Glass-Transition Temperature	193
6.13 Solvation of Linear Polymers	193
6.14 Plasticization of Cross-Linked Polymers	193
6.15 Elastomers, Fibers, and Plastics	194
6.15.1 Elastomers	194
6.15.2 Fibers	194
6.15.3 Plastics	196
6.16 Example of Polymer Physical Properties	197
6.17 Plastics	198
6.17.1 Flat Sheet Plastic	199
6.17.2 Availability and Capability	200
6.17.3 Limiting Physical Properties	200
6.17.4 Residential Plastic Materials	201
6.18 Natural and Synthetic Rubber	202
6.19 Adhesives	202
6.19.1 Cohesive Strength of Adhesives	203
6.19.2 Adherence of Adhesives	203
6.19.3 Fluidity of Liquid Adhesives	204
6.19.4 Wettability	204
6.20 Types of Adhesives	205
6.20.1 Organic Solvent-Thinned Adhesives	205
6.20.2 Latex Adhesives (Water Emulsions)	205
6.20.3 Water-Dispersed Adhesives	206
6.20.4 Two-Package Adhesives	206
6.21 Advances in Structural Adhesives	206
6.21.1 Rubber-Modified Resins	207
6.21.2 Acrylic Adhesives	207
6.21.3 Particulate-Reinforced Resins	210
6.21.4 Silane Surface Primer	210
6.22 Adhesives for Composites	210
6.22.1 Cross-Linked Adhesives	210
6.22.2 Novalak Resins	211
6.22.3 The Resol, Resitol, and Resital Stages	211
6.23 Health Hazards	211
6.24 Recycling of Wood, Paper, and Plastics	212
Terms Introduced	214
Questions and Problems	215
Suggestions for Further Reading	215

6.1 Introduction

Wood is one of our oldest building materials. In dry climates it is extremely durable. However, in humid environments wood is attacked by bacteria, fungi, and insects as part of nature's essential recycling process.

Polymers offer alternatives to wood, steel, and other building materials. To understand the potential advantages and limitations of plastics, natural and synthetic polymers are discussed in some detail together with their applications as plastics, adhesives, and elastomers.

The possible hazards of long-term exposure to chemicals are emphasized throughout the chapter.

6.2 Classification of Woods

Wood for industrial use falls into two classes: softwoods and hardwoods. Softwoods are from coniferous trees, namely, pine, spruce, fir, hemlock, cedar, redwood, and cypress. Hardwoods are dense, close-grained, and from deciduous trees, such as oak, walnut, cherry, maple, teak, and mahogany. Hardwoods have many slender elongated cells and grain irregularities that make them harder to work or split, but they provide an attractive grain pattern. The difference between softwood and hardwood is not so much in the hardness as in the degree of difficulty of working.

6.3 Composition of Wood

Trees provide the food (wood and leaves) that contains vital nutrients for many kinds of animals, including insects and microorganisms. In addition, wood is a source of numerous chemicals and building materials. Until their commercial synthesis, acetone, acetic acid, and methyl alcohol were important by-products of the destructive distillation of wood. The chemical composition of wood is complex, but the most important ingredients are cellulose, lignin, and pentosan.

6.3.1 Cellulose

Cellulose, obtained from wood and cotton, is a high-molecular-weight carbohydrate. This natural polymer is hydrolyzed into smaller molecules for industrial use by treatment with sulfurous acid and its sodium sulfite salt. Sodium carbonate, sodium sulfide, and sodium hydroxide are used as digesting agents. Chemical modifications are cellulose nitrate (known as guncotton or smokeless powder), cellulose acetate (a plastic), and methyl cellulose (a widely used thickening agent).

6.3.2 Pentosan and lignin

Pentosan, a complex carbohydrate by-product of wood, is used as an animal food. It is a natural polymer of pentose, which is a five-carbon atom of sugar.

Lignin makes up about one-fourth of the composition of wood. A by-product of the papermaking industry, lignin is a phenylpropane polymer and serves as a binder to hold the cellulose fibers together. Lignin can be separated from the wood fibers by a chemical reaction at an elevated temperature. It is used as an extender for phenolic adhesives.

6.4 Structural Panel Composites

In an effort to conserve and use the limited supply of timber more efficiently, plywood, hardboard, and particleboard panels have become well established in their use in the building industry. Plywood fills a need for thin wood panels with more structural strength and less warping than a solid sheet of lumber of similar thickness. Hardboard and particleboard panels are scrap lumber, shavings, and fibers recycled into composite panels.

6.4.1 Plywood

When hardwood is soaked with water or steamed, it can be cut into thin sheets called veneer facing. After drying and pressing flat, the sheets of hardwood veneer can be glued to enclose sheets of softwood cores. Hardwood veneer is suitable for use as paneling and as furniture.

Softwood is cut into thin veneer sheets. Three or more of these sheets are glued together to make an economical plywood with improved dimensional stability, stiffness, and strength. An odd number of thin layers is used (three to five layers), called plies. Each layer is oriented so that its grain is at right angles to the grain in the layers above and below. The strength of wood is greater along the grain than across the grain. Thus the strength of the plywood, with the grain alternating between layers, is strong in each direction. This makes high-grade plywood superior to most metals in strength-to-weight ratio (Fig. 6.1). Some of the advantages of plywood, compared to the wood it came from, are that it has a lower expansion coefficient and less tendency to warp, is stiffer and stronger, and does not split easily.

There are two common ways of grading plywood: the PS 1 system of product standards (Construction and Industry Plywood) and the ANSI-HPMA HP-1983 rating system of the Hardwood Plywood Manufacturers Association (Hardwood and Decorative Plywood). Although these standards do overlap, the PS 1 standards include dec-

Wood, Polymers, Adhesives, and Plastics   177

Figure 6.1 Plywood, attached to steel frame, covers walls, floor, and roof.

orative panels, whereas the HP-1983 standard includes mobile and modular structures. For example, PS A/D means that the front surface of the plywood is grade A, whereas the back side is a lower grade D. In the case of the HP standard, six grades of veneer are identified. Both systems of standards designate the panels for use as either interior or exterior, depending on how the glue will perform if exposed to moisture.

Performance standards are often used, especially PS 1, by the American Society for Testing and Materials to allow manufacturers to substitute ingredients. The standards emphasize the need to use quality materials to provide the margin of durability needed to meet "unusual" weather conditions.

6.4.2 Particleboard

The properties and performance of particleboard depend on the orientation of the particles, their nature, the kind of resin, and the amount of adhesive used to bond the particles together. Particleboard is known by many other names, such as composition board, chipboard, chipcore, flakeboard, waferboard, and oriented standard board (OSB).

This economical product, made by sealing a core of shredded wood chips, fibers, or particles between veneer facings, uses waste wood and small scraps of low-grade lumber. Some of the names more accurately describe the composition of the board.

6.5 Bacterial Destruction of Wood

Although bacteria are present during all stages of the deterioration of wood, their role in its destruction is considered to be minor. Bacteria are believed to act in conjunction with fungi and wood-destroying insects. This is especially true in soft-rot degradation of wood. The bacteria are unicelled and have difficulty penetrating the surface of the wood. In most cases bacterial degradation is limited to cell-wall etching and erosion in softwood regions. Although bacteria have been shown to play a part in marine pole degradation and in the papermaking processes, they normally attack wood only in especially wet environments. At most, damage occurs slowly and is small compared to the destruction caused by fungi.

6.6 Fungal Destruction of Wood

Fungi thrive if the moisture content of the wood is between 30 and 40 percent. This condition should be called "wet" rot, not dry rot, the common terminology. Dry rot is sensitive to extreme temperatures. Its growth decreases as the temperature is lowered to the freezing point. It is killed if heated to the boiling point of water. Fungi propagate best at an acid pH of 2, but unfortunately they also multiply in wood with a pH of between 3 and 5, and they grow rapidly in alkaline mortar and plaster. Outdoor dry rot is less common if there are extremes of temperature and humidity.

How does fungal infection get started? Fungi produce tiny airborne spores which, on a microscopic scale, resemble the fruit on trees. The mycelium, or root system, propagates the fungi by invading adjacent pieces of wood. If environmental conditions are right, fungal infections can also start by airborne spores falling on cardboard or other cellulosic material left in damp places. The spores can then transfer to lumber at a later date. The fungi are usually found in cellars, floors, and attics where water leaks occur from the outside. In order to occur vigorously, fungal attack requires a relative humidity of 50 percent or more.

Although some forms have large mycelium that can be seen easily, other forms digest the wood beneath the surface and may not be noticed. In general the hyphal rootlike tip of the fungus generates enzymes which break down the lignin coating and the cell wall chemically to gain access to the hemicellulose in the wood cells. To the two common types of fungal destruction of wood known as white rot and brown rot, soft rot has recently been added.

6.6.1 White rot

White rot produces enzymes which can depolymerize and metabolize all of the components in the cell wall, starting with the cell-wall surface and eating inward to get to the hemicellulose. The remaining cell wall is left intact. Lignin can also be completely degraded and metabolized by the enzymes of white rot.

6.6.2 Brown rot

The enzymes of brown rot act by chemically depolymerizing and metabolizing the structural carbohydrates in the cell walls of the wood. Lignin is degenerated but not metabolized; it remains as the residue that, along with the colored hyphae, is probably responsible for the brown color. Brown rot develops rapidly in the cell walls and results in rapid loss of strength and biomass of the wood.

6.6.3 Soft rot

The enzymes of soft rot (superficial wet rot caused by microfungi) can degrade either the cell-wall carbohydrates or the lignin, but seem to prefer the carbohydrates. High moisture content of the wood and contact with the soil seem to encourage the development of the fungi. These fungi seem to be tolerant of the high concentrations of wood preservatives in poles, pilings, and other waterlogged wooden structures and tolerant of low-oxygen conditions. Termites could be involved.

6.6.4 Rot protection

Wood differs widely in its susceptibility to fungal attack. Softwoods such as pine have the highest risk and redwood has the least risk. Most damage can be avoided by protecting the wood from continual wetting, selecting naturally resistant wood, or chemically treating the wood along with regular maintenance, especially by renewing the waterproof coating of paint.

Wood rot can be recognized visually by the coloration and structural changes in the surface of the wood. It is identified microscopically by cell-wall erosion and the appearance of the fungal hyphae and fine spores. It can be detected mechanically by inserting the sharp end of an ice pick or awl into the wood. If an effort is required to remove the sharp end from the wood, the wood is sound; if not, rot should be suspected. Rot can be noticed if the surface of the wood appears buckled or warped and can be crushed with a blunt object.

In order to thrive in wood, fungi require moisture, oxygen, a moderate temperature, and an adequate food supply. Moisture is the easiest to control. If the wood is kept dry, it should not decay.

6.7 Deterioration by Insects

No matter what we do to discourage and treat insects in our wood furnishings, homes, and commercial buildings, they are always with us. Some insects are already in the wood before it is processed. Others are lurking out there, anxious to destroy our valuable homes and furnishings. Some of these are carpenter ants, carpenter bees, and termites.

6.7.1 Wood inhabitants

Although considered pests, some wood-inhabiting insects in structures do not actually attack manufactured wood in service. They are forest inhabitants which lay eggs in the bark of recently killed trees and downed logs. Their microscopic eggs pass undetected through the manufacturing processes. These insects are killed by kiln drying, but can be built into structures where framing lumber is used without drying. Then they complete their life cycles and emerge from your favorite furnishings after as long as five years.

Ambrosia beetles, wood wasps, and two types of wood borers may emerge from expensive furniture, picture frames, and structures. Unless infestation is unusually heavy, the major damage is the defacement caused by adult emergence holes. Fumigation is sometimes considered if structural damage appears possible.

Most spiders are considered a nuisance and are usually harmless to people. They are considered beneficial since they prey upon other insects.

Widow spiders can be identified by the characteristic "hour glass" marking on their abdomen and by their unusual web. They are found in crawl spaces under homes, stored lumber, hollow tile blocks, sheds, and in unfrequented areas of closets, basements, and attics of old buildings. Indoors they may spin webs under tables and desks, in shoes, in utility ducts, and under objects in little-disturbed areas.

There is a real risk of being bitten. The bite of the brown widow spider and that of the black widow cause initial pain followed by sharp pain that may persist for several hours. The pain moves to the victim's abdomen and legs. The venom can cause nausea, dizziness, tremors, speech disturbances, and general motor paralysis. Anyone suspecting a bite by a widow spider should get immediate medical attention. Death due to venom is rare, but possible. An antivenin is available.

6.7.2 Carpenter ants

Carpenter ants are large and yellow-brown with dark-brown stripes across the top of the abdomen. They are scavengers that feed on dead insects, the honeydew excreted by aphids on trees, and crumbs and scraps found inside homes, including meat and grease. For their favorite nesting places they seek out natural cavities in wood (often what has been hollowed out by termites), decayed framing, cedar shakes or shingles, rotting logs and tree stumps in soil, and the insides of hollow-core doors, double walls, pianos, cardboard boxes, and cupboards. Carpenter ants do some excavation of wood, but usually do not destroy wooden structures. However, some species of carpenter ants in the Pacific northwest and the eastern United States are serious wood destroyers, rivaling termites in their importance.

Eradication, indoors or outdoors, requires cleaning up all food and locating and destroying the nest, not insecticidal spraying of the foragers.

6.7.3 Carpenter bees

Often mistaken for bumblebees, black carpenter bees are not social insects, as are honey bees, ants, and termites. The carpenter bee excavates larval chambers in softwoods, such as cedar, redwood, pine, exposed rafters, old lumber, telephone poles, fences, and picnic tables, to create entrance holes and tunnels about 1.27 centimeters (½ inch) in diameter and 5 to 20 centimeters (2 to 8 inches) in length (Fig. 6.2). The tunnels run with the grain and consist of a series of chambers where one egg is laid over a ball of pollen, then sealed in until the larva hatches in two or three days. Only one female will collect pollen, prepare cells, and lay eggs. Other females guard and clean the nest. Abandoned galleries are reused, and their length and complexity can be greatly increased over time. A wooden structure may become useless if this excavation is allowed to continue.

Control requires dusting the entrance holes with an EPA-approved insecticide. The dust should be puffed into entrances, preferably at

Figure 6.2 Carpenter bees tunnel far into redwood beams and eaves, threatening the strength of the wood.

sunset, and the bees should be allowed to pass freely over the treated area. This allows the adult bees to transport the insecticide into the galleries. New emerging bees will brush against the insecticide. Inaccessible entrances, such as those high up on the eaves, can be treated with an aerosol stream. Painted wood is reported to be less susceptible to carpenter bee damage.

6.7.4 Termites

The West Indian termite, probably introduced from tropical America before 1869, is the most destructive of the three species of drywood termite. It feeds on most types of wood found inside and outside the home, preferring hardwoods, which are often used for flooring and expensive furniture. Also, it attacks books and other cellulose-containing materials. Swarms usually occur on warm summer nights and are attracted by nearby street or house lights. They look for suitable sites for new colonies: cracks, nail holes, and wooden joints (Fig. 6.3). The water they need comes from normal humidity and the digestion of food. This termite, dark brown in color, is often confused with the Formosan termite, which is amber or honey-colored.

Figure 6.3 Structural beam severely damaged by termites.

The Formosan subterranean termite, commonly called the "ground" termite in Hawaii, has been introduced along the United States Gulf Coast. Nests may number as many as 10 million individual termites. Because of the large population and their aggressive nature, they do extensive damage in a few short years. They are known to tunnel through wood foundations and cement. A conservative estimate of the damage in Hawaii alone is said to exceed $60 million a year.

There may be little or no external evidence of attack. The first indication of infestation may be a sagging floor, leaking roof, warped walls, hollowed-out beams, depressions in wood, blistered paint, a short-circuited electric circuit, or similar problems. Unprotected homes, that is, homes built of untreated lumber and homes built over strong colonies, have been almost completely destroyed within 2 years.

Swarming usually occurs on calm, humid summer nights. Thousands of swarmers are attracted to the closest light source. The few that survive predators (lizards, spiders, toads, or ants) must find the right conditions (food, moisture, and shelter) to start a new colony and continue the damage.

The termites' food, cellulose, is found in the walls of all plants, such as wood, paper, fruit, nuts, cork, or living plants. The cellulose is broken down by protozoans or bacteria living in the termites' intestines; otherwise termites would starve to death.

The moisture requirement can be met by high humidity. Free water is not required. Moisture can come from damp soil, leaking roofs or plumbing, air conditioners, pipes, or poorly designed decks and roofs.

The shelter requirement can be met by a joint, nail hole, crack, or other crevice that can be entered and sealed to form a new home.

One pair can start a colony within the sealed chamber, with the first batch of eggs hatching about 30 days later. This process continues until a major colony of 2 million or more termites is produced, which takes more than 7 years. Even if cut off from ground contact, much damage can be done to the wood by the remaining termites. If infestations are not spot-treated vigorously, fumigation will be required.

6.8 Prevention and Control of Termites

Two basic strategies in the management of termites are prevention and remedial control. Prevention involves good architectural design, the use of nonfood items in critical areas, the use of treated wood, and the placement of a barrier in the soil between the termite colony and the structure.

Preventive measures include using pressure-treated wood containing chromated copper arsenate (CCA) and Osmose and Wolman pressure-treated woods. Treated woods may be effective against Formosan and West Indian adults seeking new territory for a colony. CCA-treated wood cannot withstand a sustained attack by Formosan termites because of insufficient penetration of the ingredients into the heartwood. Recent tests conducted by the termite research laboratory at the University of Hawaii–Manoa, Honolulu, revealed that ammoniated copper zinc arsenate (ACZA), with its superior penetration into the wood, is most effective against the Formosan termite. Wood, pressure-treated with ACZA, was untouched and killed the Formosan termites.

Control of termites requires a careful, well-planned immediate attack. When swarming begins, turn off lights to reduce the attraction. If swarmers are inside the building, look for flight slits inside. Kill tandem pairs running around. Periodically inspect in and around the building. Keep building walls clear of plants, which provide food and hide tunnels. Avoid having scrap wood touching the ground around the premises. Inspect new materials brought in to be sure they are termite-free. Break the food, moisture, and home requirements—any one of which can stop the termite.

Treatment, after infestation is discovered, requires a careful, well-planned immediate attack. Long-lasting ground fumigants, such as aldrin, chlordane, and heptachlor, have been taken off the market because of their human health hazards. Products on the market that are EPA-approved can slow down the growth of colonies temporarily. Another method, fumigation, is discussed later in the chapter.

6.8.1 Basaltic termite barrier

The basaltic termite barrier, a nontoxic method of preventing infestation by the subterranean termite, has been developed by researchers at the University of Hawaii–Manoa, Honolulu. A permanent physical barrier of basaltic aggregate is formed between the ground and the foundation of the house or other structure. The basalt can also be used to form a barrier around the perimeter of footings of older construction.

The spaces between the granules of basalt are too small for the termites to crawl between, and if a granule is dislodged, other granules slip in to close any openings for tunnels. The material is too hard for termites to chew (a specific gravity of 2.9) and too heavy to move. The optimum granule size is between 1.7 and 2.0 millimeters. Shape and density requirements make the basalt-sieved particles most suitable.

Proper construction techniques and maintenance are required, even though the barrier appears to be the best technical solution against ground termites. This environmentally acceptable barrier of basalt does not need to be replaced and outlasts chemical barriers.

6.8.2 Cell treatment of wood

The most popular wood in general use is the green-colored lumber that is chemically treated to resist rot and insects. Almost all framing is done with this lumber. Treatment is called full-cell and empty-cell, although neither condition is fully achieved.

In full-cell treatment, pressure is applied to force the preventive fluid into the cellular structure of the wood. In empty-cell treatment, pressure is used to wet the cellular walls. Both methods require large containers that can be pressurized to 10 or 15 times atmospheric pressure. The process is expensive and requires the wood to be shipped to special treatment centers.

6.8.3 Protection by fumigation

Tenting is one way of getting rid of termites and other insects already infesting a building. This involves wrapping a house tightly with a plastic-laminated canvas, pumping sulfuryl fluoride (SO_2F_2), a toxic gas, into the house, circulating the chemicals with fans, and leaving the house sealed for 24 to 72 hours. The treatment lasts for 3 to 4 years, costs about $700 to $1000 per house, and has little residual effect. Only the live insects are killed and, perhaps, the eggs. It does not prevent future infestations.

6.9 Paper Technology

Long-lasting paper was made by hand from flax (linen), cotton, and natural grasses until the midnineteenth century. Development of a continuous mechanical process of grinding wood into pulp modernized the papermaking industry. However, the wood fibers are shortened and crushed in the grinding process. Even the paper made from softwood, with its longer fibers, has little strength and flexibility. Use of the shorter hardwood fibers further shortens the useful life of paper. In the presence of sunlight, moisture, and oxygen from the air, the life of newsprint is about 2 weeks. However, under ideal storage conditions, its life may be extended to 20 years.

6.9.1 Modern papermaking

The papermaking industry was revolutionized in the middle of the nineteenth century by the soda, the sulfite, and the sulfate (or kraft) processes. These processes are in use today in modified form. Use of the less expensive acid sulfite processed paper has become widespread, not only in industry, but also for books, magazines, and other documents.

The cellulose usable for papermaking comprises only about half of the wood, with lignin, pentosans, and terpenes making up most of the remainder. In the chemical digesting processes the wood chips are cooked for several hours at 150°C (302°F) with superheated steam. Soda ash (sodium carbonate) is the alkaline-digesting chemical in the soda process. In the kraft process the chips are digested in a sodium hydroxide, sodium sulfide liquor. In another process the wood chips are cooked in sodium or calcium bisulfite.

In the preparation of the paper, the cellulose fibers are filtered from the digested wood pulp, washed, bleached with a chlorine compound, and sized with an alum-rosin mixture. Finally, fillers such as calcium sulfate, clay, and kaolin are added for opacity and ease in printing. To make shorter fibers, the hydrolysis of cellulose is catalyzed by acids. Alums, especially papermaker's alum (aluminum sulfate), hydrolyze in water to form sulfuric acid. Hydrochloric acid is formed by the reaction of chlorine with water.

6.9.2 Self-destruction of modern paper

Unfortunately, we are discovering that documents prepared after about 1850 are starting to crumble and disintegrate. The relatively short life of our present paper threatens to cause an enormous gap in the recorded history of modern civilization. The culprit appears to be the acid left in the finished paper. This acid comes from numerous sources, including the difficulty of washing the acid and acidic salts from the pulp, from the bleaching process, from the alum-rosin sizing, and from acid added to the printing ink. All of this is in addition to the effects of air pollutants.

Although other destructive factors are involved in making paper, the residual acid in the paper has been the chief concern in prolonging the life of our manuscripts. The crisis of self-deterioration of paper is a challenge to the paper industry and to libraries throughout the world.

6.9.3 Treatment of acid paper

To delay the deterioration, records can be stored in libraries at a temperature below 18°C (65°F). Or their valued paper contents can be treated to neutralize the residual acid. The methods being tried are (1) dipping each book in a suspension of magnesium oxide in freon, and (2) bathing books in diethyl zinc for 8 hours. In the meantime the papermakers are beginning to produce paper to acid-free specifications that includes a pH of at least 7.5.

One recommended precaution: store paper records in air-conditioned libraries with controlled temperature and humidity. Since accelerated testing is, at best, only a guide, there is wide diversity in the technologies developed to neutralize the acid in paper manuscripts. The magnitude of the problem is immense because all books and records printed in the last 150 years may be in danger.

6.10 Polymers

Cellulose nitrate (celluloid), one of the first synthetic plastics, was made by treating cellulose fibers (from cotton or wood) with nitric acid. Plasticized with camphor to make it flexible, it was used as celluloid windows in the early automobiles. But cellulose became brittle as the camphor escaped, and sunlight caused it to decompose and turn yellow. This modification of a natural polymer was followed in 1909 by the commercial production of Bakelite from phenol and formaldehyde.

Polymers are now readily made with their molecular weights ranging from a few thousand to several million. Although their chemical properties are largely determined by the monomer, their physical properties depend mainly on the length of the polymer chains and their structural arrangement.

6.10.1 Natural and synthetic polymers

Polymers occur abundantly in nature as proteins and carbohydrates. The protein polymers include keratin, the fibrous constituent of hair, feathers, hoofs, and horns. Collagen, making up most of the fibrous protein in the body, also occurs in the rigid, strong bundles of fibers in bones. Vegetable fibers include complex carbohydrates as well as the polysaccharide sugars, pectin, and starches. Textiles have long been made from cotton, silk, and wool fibers. The natural cellulose polymer plays a predominant part in the structure of wood as fibrous bundles of protein molecules in a polyphenol.

Natural polymers are unique because of the complex structure of their macromolecules. Proteinlike enzyme catalysts speed up reactions in animal and plant cells. These enzymes also direct the orderly placement of monomer units to provide the strength needed in sinews and organs of the body. Other enzymes help build the fibrous structure of wood. Many of these enzymatic engineered structures still challenge the polymer chemist.

In view of the varied properties of polymers that are synthesized by enzymes, consider the progress made by polymer chemists in duplicating the achievements of nature. It is an axiom of science that before research can begin in a new area, instruments must be available to identify and study it. Recent developments in expensive, state-of-the-art research instruments are producing exciting results.

6.10.2 Structure of linear polymeric chains

As the structures of polymers become better understood, it is possible to synthesize macromolecules in unique types of ordered carbon chains. To understand the effect of structure on physical properties, it is necessary to have a more detailed picture of linear polymers.

Polymer chains become flexible enough to overlap, tangle, and bond to one another if they exceed a critical length of about 2000 carbon atoms or 1000 monomers. The number of monomers in a polymer is known as the degree of polymerization, indicated by DP or n. But what do polymer chains look like?

Consider the shape of a polyethylene (PE) polymer chain made by the ordinary free-radical polymerization process. Instead of being a single strand of a spaghettilike chain of carbon atoms, it has methyl side groups attached to it at random. The polymer has what is known as an atactic structure. These methyl side groups keep the chains in the polymer from sticking together, thus allowing individual chains to move more easily over one another. This ease of chain movement causes the polymer to be more easily deformed.

Kitchen utensils made of ordinary polyethylene do not hold their shape in the hot water used in a dishwasher. However, by using a special anionic catalyst mounted on a crystalline surface, the methyl side groups can be located in an orderly sequence so the chains can fit snugly together. The effect of the many weak bonds formed between the atoms produces regions of rigidity and a semicrystalline solid. The result is a high-density polyethylene (HDPE) solid. This partially crystalline solid requires a higher temperature to be reached before it softens. Thus high-density polyethylene retains its shape in boiling water.

Further research taught chemists how to alternate the methyl side groups from one side of a polymer chain to the other to form a syndiotactic structure. They also learned how to get the methyl side groups to attach only on one side of the chain to form what is called an isotactic structure. The symmetry of these types of structures can be obtained with other linear polymers, such as polypropylene and butyl rubber. Copolymers offer additional structural possibilities.

Linear copolymers made from two different monomers usually result in the monomers occurring in a random order in the chain. Polymers can also be made in a symmetrical, alternating sequence along the chain. Also, copolymers can be formed with blocks of one monomer alternating regularly with blocks of the other monomer. The first type of a random block polymer structure would be an amorphous, or noncrystalline, solid. The copolymer with alternating block symmetry produces crystalline structures with higher melting points and structural strength. Note that these changes in physical properties have been obtained without any change in the chemical composition of the polymer.

Among the advantages of polymers with regulated chain structures is the combination of hardness and toughness, which usually do not occur together. Ordinarily, a hard substance is brittle. A diamond, for instance, is highest on the hardness scale. This is in contrast with soft graphite, an allotropic form of carbon. Yet in these special polymers we have toughness without brittleness—a useful advantage in fibers and other applications.

6.11 Physical Properties of Polymers

Polymers can be made easily to have unique physical properties. These properties include a characteristic glass-transition temperature, plasticity when warmed, a large thermal expansion compared to metals, viscoelasticity (illustrated in Figs. 6.4 and 6.5), and high permeability to some vapors and gases.

6.11.1 Permeability of polymers

Because of the thermal motion of polymer segments at ambient temperature, the regions around the functional groups of the polymer are permeable to gases and water vapor. Both natural and synthetic

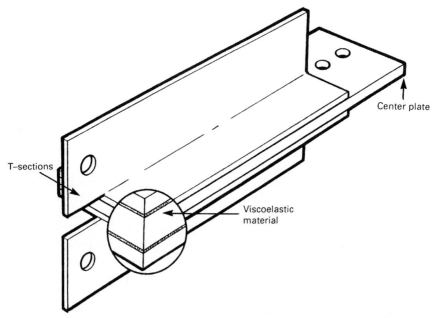

Figure 6.4 Dampers of viscoelastic material bonded to steel plates are used in locations that will experience relative motion during a wind or seismic event. They are designed to help dissipate vibrational energy to which the structure may be subjected. (*Courtesy of 3M.*)

Wood, Polymers, Adhesives, and Plastics 191

Figure 6.5 Viscoelastic damping materials bonded to steel plates and girders damp building vibration for improved human comfort at 2 Union Square, Seattle, Wash. The energy of vibration, caused by strong winds and seismic movements, is dissipated as heat through the steel members of the numerous dampers. (*Courtesy of 3M.*)

materials, from a loaf of bread to Styrofoam, are permeable to carbon dioxide. The naturally occurring carbon dioxide in bread escapes faster than air can replace it, so bread shrinks as it ages.

Styrofoam, more permeable to carbon dioxide than to air, is stored for a week or more before use to avoid a change in shape when put into use. Thus the solubility of gas molecules in the polymer is more important than their molecular weight in these examples.

6.11.2 Coefficient of expansion

Organic polymers have larger thermal expansion coefficients than metals or other building materials. The thermal expansion coefficients of linear polymers are about twice those of cross-linked polymers, which in turn are about twice those of aluminum or glass. To better match the expansion of polymers as they warm, inorganic fillers are used. Most common fillers are powdered silica, aluminum, and aluminum oxide. Besides providing a better match of expansion coefficients, fillers can provide corrosion resistance, shear and tensile strength, and reduction in gas and vapor permeability.

6.11.3 Glass-transition temperature

As polymers are heated, the thermal motion of the chain segments increases until their kinetic energy overcomes the weak bonding energy, causing the polymer to soften and become plastic. This temperature is called the glass-transition temperature. The polymer goes from a glasslike solid to a soft pliable mass. Only if heated to a higher temperature will the polymer be fluid enough to be considered melted. This gradual melting property of polymers is caused by the tangle of cross-linked chains. Glass also has this property, as the name implies.

6.11.4 Lowering the glass-transition temperature

Plasticizers should match the polar nature of the polymer and should not be volatile. There are two types of plasticizers that can effectively lower the glass-transition temperature of a polymer. Reactive plasticizers can be copolymers, which also reduce the attractive force between the polymer segments. The unreactive plasticizers are high-boiling-point solvents. Dibutyl phthalate has a boiling point of 340°C (644°F) and can make up close to 30 percent of the volume in the commercial polymethyl methacrylate plastic. The purpose of the plasticizer is to make the product less brittle and easier to form into tubing and sheets.

6.12 Example of Glass-Transition Temperature

Directions. Gradually warm a thin strip of Plexiglass by holding it over a hot plate or passing it back and forth, cautiously, through the flame of a gas burner until the glass-transition temperature is reached (as shown by the flexibility of the plastic strip). *Note:* To avoid burns, it is best to wear cloth gloves when handling hot objects.

Wind the hot strip into a figure, such as a spiral, and submerge it into cold water for a few minutes to cool it below the glass-transition temperature and to stiffen the strip. Then, to show that the tangled polymer chains have a memory, gently warm the spiral until it uncoils. The chain segments restore their weak bonds to one another to resume their previous tangled arrangement.

Interpretation. The glass-transition temperature of polymethyl methacrylate is at about the boiling point of water. (The transition temperature depends on the amount of plasticizer used.) The poor thermal conductivity of a strip of plastic, relative to metals, is shown when you are able to hold one end of the strip of plastic while the other end is held in a flame a few inches away. If time permits, put a strip of plastic and a thermometer in a beaker of water and, as the water is heated, measure the temperature when the strip becomes rubbery, showing that the glass-transition temperature is reached.

6.13 Solvation of Linear Polymers

Pockets between the chain segments provide spaces into which solvents can penetrate. Although solvent molecules can break some of the weaker bonds and separate some of the chain tangles, especially if warmed, only a partial solution of chunks of polymer chains is possible if the polymer has a large molecular weight.

6.14 Plasticization of Cross-Linked Polymers

If the monomer has more than two functional groups, forked chains and cross-linkages will result in a thermosetting polymer. The attraction of the solvent for this type of polymer is insufficient to break the chemical bonds and form a dispersion, and it is only able to produce a swollen, rubbery solid.

6.15 Elastomers, Fibers, and Plastics

Some monomers and their combinations produce polymers with the physical properties needed for elastomers, fibers, or plastics more easily than others. Such properties can be built into almost any polymer.

6.15.1 Elastomers

To be elastomeric, or rubbery, a polymer, when stretched, must be able to break the weak interchain tangles that make it amorphous such that the randomly oriented chains can take more extended positions, yet will return to disordered tangles when released. To prevent the flow or gliding of chains over one another, resulting in a permanent deformation, some cross-linkage between chains is usually necessary.

Two types of cross-linkage, hydrogen and covalent bonding, are usually involved between the chains of an elastomeric polymer. The crystal structure of the chains is more permanent if covalent chemical bonding occurs. A certain amount of bonding between chains is necessary to prevent permanent elongation when stretched. The weaker hydrogen bonds formed between hydrogen and the oxygen or nitrogen atoms attached to the chains are readily broken by stretching, heating, or by active (polar) solvents. These bonds are restored by the reverse treatment.

Elastomers can be prepared from synthetic polymers such as polyurethanes, polyesters, and polyethers by block polymerization with the formation of alternating polar and nonpolar sections in the chains to provide for crystalline regularity. These can be plasticized with a nonvolatile solvent as needed.

6.15.2 Fibers

For a polymer to have both the tensile strength and the flexibility required of a fiber, the polymer chains need to have an ordered structure that allows crystallizing forces to hold bundles of the chains firmly together. Cold stretching is a common practice after the extrusion of fibers to pull the kinks out of the chains, align them, and bring them into closer contact. It is usually undesirable to have excessive cross-linkage between chains as this would reduce the flexibility of the fiber.

One important use of fibers is in carpets and textiles. Tufted carpets, in contrast to the more inexpensive knitted and woven types, are made by stitching the pile to a prefabricated jute backing. Punched or flocked outdoor carpet has an elastomeric coating on the bottom to hold the polypropylene or nylon pile in place. An advantage of these synthetic fibers over wool is the lack of nutrients for moths and fungi.

Carpet fibers wear better if packed closely together so that the ends, rather than the sides, of the fiber are exposed to wear. Consequently the design of the carpet is important as well as the type of fiber used. The compactness of the fibers and how tightly they are secured to the backing can be observed by flexing a carpet.

If the carpet does not have its own built-in padding, the life of the carpet can be increased with the selection of a resilient pad. Its use reduces uneven wear from an irregular floor, rubbing on the floor, and penetration of dirt, and it provides thermal insulation.

Carpet fibers can be natural or synthetic. Synthetic fibers, including regenerated cellulose usually known as viscose rayon, are formed by extrusion through a small orifice in a continuous filament. These are chopped into lengths to give the appearance of wool.

Blending of different fibers is a common practice, such as the addition of nylon to improve the wear of viscose rayon. Also known as modified or cellulosic rayon, viscose rayon is made from cotton or wood chips that have been solubilized (plasticized) by treatment with caustic soda, reacted with carbon disulfide, and extruded into an acid neutralizing solution to obtain either cellophane sheets or fibers of viscose rayon.

Cellulose acetate and cellulose nitrate are other chemical modifications of cellulose. They are also made from cotton or wood. The search for polymer fibers to replace or improve upon natural fibers has led to the development of many varieties of promising synthetic polymers.

Nylon is a good illustration of the attempts to improve on nature (see Fig. 6.6). Although at least six or seven nylons are commercially available, they differ little in physical properties. Made from different patented processes and varied raw materials, they are all polymers of the polyamide thermoplastic crystalline type, with nylon 66 and nylon 6 the most frequently used.

Nylon 66, patented in 1934 as a substitute for silk, is made from two six-carbon-atom monomers called adipic acid and hexamethylenediamine. Nylon 6 is the polymer of a six-carbon-atom monomer with the acid radical on one end and the amine on the other. Nylon 66 has a melting point of 264°C (507°F) compared to 223°C (433°F) for nylon 6.

Figure 6.6 Brushes with polyester/nylon filaments. Designed to resemble Chinese bristles in their performance, these roughened plastic filaments maximize the amount of paint delivered to a surface. (*Courtesy of Dunn-Edwards Brushes.*)

Polyester, polypropylene, and poly(vinyl chloride) are popular fibers. Polyester textiles have the advantage of holding a crease when pressed because of their relatively higher glass-transition temperature. High-density polypropylene articles hold their shape at a high dishwasher temperature for the same reason. Poly(vinyl chloride), useful as a copolymer, is a flame retardant.

Fluffy or finely woven synthetic textiles are a serious fire hazard due to the abundance of oxygen from air in the porous weave. As combustion can occur only at the surface, its rate is proportional to the area of the surface exposed. These small synthetic fibers create another hazard if they are thermoplastic because they can melt and stick to the body.

Poly(vinyl chloride) releases chlorine gas when it is burning; this gas suffocates the flame. Fire-retardant ingredients that may be incorporated into the polymer include borates, carbonates, or phosphates, which decompose to form a nonflammable gas that can smother a flame.

6.15.3 Plastics

Plastics are organic compounds that, in some stage of formation, can be shaped by flow and can be molded. Their structure and high molecular weight give them unusual properties. Plastics need to have a compromise of fibrous and elastomeric properties. To be flexible instead of brittle like glass, the polymer must be used above its glass-transition temperature. This temperature is more dependent on chain structure, freedom of chain movement, stiffness, and interchain bonding than it is on molecular weight. For example, polystyrene, polymethylmethacrylate, and poly(vinyl chloride) are brittle and have a low impact strength at ambient temperatures. This can be overcome by lowering the glass-transition temperature when adding a plasticizer to the polymer. Low-volatility liquid solvents that are commonly used as plasticizers include dibutyl phthalate and dioctyl phthalate. The plasticizer tricresyl phosphate also acts as a fire retardant.

Unfortunately, plasticizers slowly exude from polymers, evaporate from the surface, or dissolve in detergents or other cleaning fluids. The product eventually becomes brittle again. In addition, these plasticizers are toxic aromatic substances and usually carcinogenic. Longtime exposure to low concentrations of toxic vapors and carcinogens are of increasing concern to the public.

Injection molding is a process of forcing a heated fluid plastic into a mold. On cooling, the plastic solidifies into the shape of the mold. Thermoplastic polymers are suitable because they soften when heated. But thermosetting polymers can also be injection molded if they are only partially polymerized. Partially cross-linked, they are still plastic enough to be injection molded. Bakelite is an example of a thermosetting plastic which, if not fully cured, will soften on heating and can be injection molded. The cross-linking process continues during the injection molding.

6.16 Example of Polymer Physical Properties

Flammability test. The response of textiles to heat is characteristic of the polymer and its structure. When exposed to a flame, a thermoplastic fiber will melt and shrink away from the heat; a thermosetting type will char and extinguish itself by insulating itself from the flame.

First take the precaution to fill a pan with water into which any demonstration that burns too vigorously can be immersed and extinguished. Always remember that doubling the quantity of materials in an experiment may increase the danger of it getting out of control tenfold, not twofold.

Hold a small strip of carpet or textile made of fibers of wool, cotton, nylon, polyacrylonitrile, polyester, polypropylene, or polyurethane with metal tongs or a long-handled forceps. Put the sample into the edge of a gas flame. (A bunsen or laboratory gas burner is convenient, but a portable butane torch can be used.) *Caution:* A severe heat burn can result if the fibers melt and drip on your skin.

Interpretation. Wool burns slowly unless it is fluffy and contains a considerable amount of air. It leaves a charred residue, which tends to extinguish the flame. The char can be brushed off after it has cooled. Cotton or rayon ignite more easily than wool, especially if fluffy, and burn rapidly. By contrast, polyamide fibers such as nylon, Dacron polyester, and acrylonitrile will soften and draw up into a compact mass (due to surface tension) when melted.

6.17 Plastics

Plastics are an important part of our lives, replacing metals, glass, ceramics, and other common building materials. Indeed, the volume used probably exceeds that of metals (Table 6.1).

Sheet plastic is an important illustration of the replacement of glass where a nearly unbreakable, shatterproof, transparent lightweight material is needed. Such uses include glazing of windows and skylights with optional features of solar and glare control. Exterior uses include enclosures of elevated walkways between buildings. Interior uses include display windows, curtain walls, and space dividers.

The physical and chemical properties of a plastic depend on the polymer used. Continual advances in the science and technology of plastics have resulted in sturdy, durable lightweight materials, carefully engineered for architectural uses.

TABLE 6.1 Plastic as a Building Material

Type of plastic	Applications
Acrylic	Glazing, lighting fixtures
Acrylonitrile	Window frames
Polybutylene terephthalate	Countertops, sinks
Polycarbonate	Flat sheets, windows, skylights
Polyethylene	Piping
Polyphenylene oxide	Roofing panels
Polystyrene	Insulation, sheathing
Polyurethane	Insulation, roofing systems
Poly(vinyl chloride)	Molding, siding, window frames
Urea formaldehyde	Countertops

Because of the relatively low stiffness of plastics, they are not used as primary load-bearing structural materials, but they are used for transmitting loads to other structural components. These plastic materials rarely fail, but when they do, it is usually caused by the designers' or installers' lack of familiarity with their capabilities and limitations.

The critical difference between plastics and structural metals is the mechanism by which they respond to load. Plastics act differently than other building materials under dynamic (fluctuating) load. The factors which determine static load performance, such as rigidity, strength, and elastic recovery from deformation, show that plastics are significantly lower in strength and stiffness. It is most important to realize that plastics require different test methods, different interpretation of the conventional data, and a different approach to the generation and use of design data.

The selection and use of plastics, being mindful of their different kind of strength and rigidity, requires a different approach than the one used with structural metals because of their viscoelastic properties. The "perfect spring" model used for metals must be replaced by a complex model combining features of a semicrystalline solid and a viscous fluid. In other words, due to the temperature of creep, both the strength and the stiffness of plastics are time-dependent.

Plastics consist of one or more high polymers with additives, such as a plasticizer, which in some stage of formation are fluid and can be spread or molded before solidifying. Their structure and high molecular weight give them unique properties. To modify a brittle plastic so that it becomes more tough and flexible, its glass-transition temperature (above which it becomes plastic, like hot glass) must be lowered by the addition of a plasticizer. The plasticizer increases the freedom of polymer-chain movement and interchain bonding.

6.17.1 Flat sheet plastic

Sheets of acrylic or polycarbonate plastic are made by cell casting or by a continuous process. In cell casting, optical-quality sheets are made individually one at a time in polished plate-glass molds. By changing the thickness of the vinyl spacer, any thickness can be custom-made from 0.15 to 10 centimeters (0.060 to about 4 inches) in thickness. The sizes range from 30 by 90 centimeters (2 by 3 feet) to 300 by 360 centimeters (10 by 12 feet). Patterned sheets are limited in size to that of the available patterned plate glass.

Tinted, mirrorized, and hollow-core acrylic sheets are also made in limited sizes and thicknesses. Tinted and mirrorized sheets are made by vacuum evaporation of metals on the back surface of the sheet. Hollow-core (double-skinned) sheets have ribs to separate the front and back sheets. These lightweight polycarbonate sheets find many unusual applications because they are practically unbreakable. In fact this feature has raised concerns by municipal fire departments that in case of an emergency, they might not be able to break through a polycarbonate door to put out a fire.

Continuous cast-sheet plastic is made by pouring a catalyzed liquid monomer onto a continuously moving stainless-sheet belt and polymerizing the plastic as it passes through an oven. This continuous sheet is rolled up around a reel. After reaching its destination, the roll of plastic sheet is heated, unrolled, and reshaped as needed. The length of plastic is from 75 to 180 meters (250 to 600 feet) in widths of 0.9 to 2.4 meters (3 to 8 feet). The length of continuous flat sheets is limited by the length that can be transported to the work site.

6.17.2 Availability and capability

Acrylics and polycarbonates are unique building materials because of their long-term resistance to weather without a significant deterioration in appearance or properties. Polycarbonate has phenomenal impact resistance, which remains higher than that of many other transparent plastics such as poly(vinyl chloride), cellulose acetate butyrate, and polystyrene.

Large sheets, shades of tint, and grades of materials offer reduced reflection, increased opacity, and abrasion-resistant surfaces.

Heated plastic can be poured into a mold. On cooling, the plastic solidifies into the shape of the mold. Thermoplastic polymers are suitable because they soften when heated. But thermosetting polymers can also be injection molded if they are only partially polymerized. Partially cross-linked, they are still plastic enough to be injection molded. Bakelite is an example of a thermosetting plastic which, if not fully cured, will soften on heating and can be injection molded. The cross-linking process continues during the injection molding.

6.17.3 Limiting physical properties

The effect of temperature changes on physical properties is different for plastics than for other building materials. The thermal expansion and contraction are much larger for plastics than for most other materials. Also, the glass-transition temperature of plastics is much closer to ambient temperatures than it is for other building materials. This is due to the unique nature of the molecular bonding and structure of polymers. In contrast to plastics, glass has a glass-transition temperature that is safely above ambient temperature to influence its physical properties when used as a building material.

As a consequence of their large thermal expansion, plastic sheets curve when the surface that is exposed to the sun becomes hotter and expands more than the back surface. There is also a tendency for a plastic sheet to sag under its own weight if it is mounted at an angle to the vertical. Polycarbonate plastic sheets provide greater stiffness, toughness, and impact strength than acrylic sheets.

The hardness of plastics is difficult to test because they have a different structure than other building materials. Special tests are needed to measure their physical properties such as hardness. For example, the scratch test used to measure the hardness of crystalline materials is completely unsuitable for plastics. When a scratch is made on a plastic surface, within seconds the scratch starts to fill in. Unless the surface is badly distorted, or foreign material is embedded in the process, the surface can be seen (by using a grazing light and a low-powered magnifying lens) to immediately start flowing back together and virtually healing itself. Other well-established tests give ridiculous results when used to measure hardness and other physical properties because of the plastic properties.

All plastics are combustible if heated above their decomposition temperature. Some plastics, such as poly(vinyl chloride), are self-extinguishing when the flame is removed due to gases formed in the decomposition process that act to smother the flame. It must be remembered that at the high temperatures reached in a fire, wood burns, glass shatters, concrete is unable to maintain its strength and crumbles, and metals lose their structural strength.

6.17.4 Residential plastic materials

A model house has been designed by the General Electric Company to use the maximum amount of plastics. It contains about 30 percent, or 20,000 kilograms (44,000 pounds), of plastic compared to an estimated 5 percent of plastic ordinarily used in a home. This model home has a plastic roof. (Roofing sheets look like cedar shingles, but are made of polyphenylene oxide resin.) Plastic siding on the model home is made of an acrylic styrene acrylonitrile laminated over acrylonitrile styrene butadiene. Windows, made of polycarbonate plastic, are used for maximum security against burglars. Other special features include plastic sinks, drainboards, and countertops of polybutylene terephthalate.

6.18 Natural and Synthetic Rubber

Natural and synthetic rubbers are examples of elastomeric substances that recover fully when stretched to twice their length. Examples are butyl, neoprene, nitrile, and polysulfide rubbers. They are used to modify thermosetting polymers, improving their resistance to peel and fatigue. It is necessary that their glass-transition temperatures be below room temperature for them to be elastomeric.

Natural rubber is a linear polymer of isoprene obtained as latex from the rubber tree. The polymer chains have a pair of conjugated double bonds that can be cross-linked with sulfur, which acts as a vulcanizing agent.

Synthetic rubbers include polybutadiene, made from the monomer butadiene, and a chlorinated derivative of butadiene, called chloroprene, which is used to make neoprene rubber. Both of these monomers have conjugated double bonds, that is, two double bonds separated by a single bond.

Isoprene has a methyl side group on every fourth carbon atom of the chain. Chloroprene has a chlorine atom in this position. To be elastomeric, these polymers need to have a semicrystalline character with each of the side groups on the same side of the chain called a cis structure. It would be trans if the groups were trans, or across, from one another. In the vulcanization of any of these rubbers, only partial chain cross-linkage between the extra double bonds is needed to restore the structure to the unstressed shape.

Retreading of tires and "reclaiming" of rubber are possible because of residual double bonds. Old rubber tires can be plasticized by reducing the molecular weight and chain length with heat energy from the mixing process. Some of the remaining double bonds can then be vulcanized with additional sulfur. Ozone causes the residual double bonds in rubber to cross-link at the expense of its rubbery nature. The presence of ozone in photochemical smog is shown by the hardening of erasers on the ends of pencils and by the aging of rubber bands.

6.19 Adhesives

The development of synthetic resins with superior properties has resulted in the increased use of adhesives in the construction industry. Adhesives have an advantage over rivets and bolt fasteners by distributing stress over larger areas of a joint. This reduces galvanic corrosion between dissimilar metals and provides the ability to cement together extremely thin sheets. The disadvantages of adhesives include their poor resistance to peel, the need to clean surfaces thoroughly, care in the application of the adhesive, and the time needed for the bond to form. Solid structural adhesives must first be plasticized by the use of heat or solvents. Solvent-type adhesives are limited in use to porous materials which allow the solvent to escape as the bond forms.

The important physical properties of adhesives are cohesive strength, adherence, fluidity, and wettability of the substrate.

6.19.1 Cohesive strength of adhesives

In the design of an adhesive system the adhesive should have more cohesive strength than either of the surfaces being held together. Usually the adhesive becomes more brittle as its cohesive strength is increased, thus imposing a practical limit on the improvement in strength for an adhesive. Creep, or the tendency to flow under tension, is useful in spreading the strain over a larger area, but can decrease the strength of the bond.

6.19.2 Adherence of adhesives

For an adhesive to hold two surfaces together, two conditions must be met: there must be the formation of a strong bond at each of the interfaces between the surface and the adhesive, and the adhesive must have strong cohesive strength. Since adhesive failure occurs in the weakest bond, failure can occur at either interface, and cohesive failure may occur in either surface. Or the failure may occur inside the adhesive itself. In practice, bond failure rarely occurs at an interface.

Bond failure is more apt to start at an oxide defect in the surface, at an imperfection in the adhesive, or at a microscopic clump of soil missed in the surface preparation. Thus close inspection of a joint failure will usually show a mixture of adhesive and cohesive failure.

Consequently, the authors have proposed the use of the term "hesion" as more appropriate. This is especially useful in reporting research results on bond failure. Since the use of the term adhesion is so well established, a practical way to describe the failure of a coating would be to report the percentage of adhesion and cohesive failure. The percentages of failure could be determined by using a metallographic microscope.

Because of the increased chance of a flaw, vapor pocket, or other potential points of failure occurring, only enough adhesive should be used to prevent the two surfaces from being in contact. For good hesion, not only must the adhesive be selected to have adequate tensile strength, but it must form strong chemical bonds to clean and wettable surfaces.

Surface cleanliness is essential for an adhesive to form a strong bond. Most dirty surfaces are covered with clumps of oily film which separate the adhesive from the surface. Since the film has little strength, its presence produces an area of low hesion. The reason is that most dirty surfaces have an oily or hydrophobic water-repelling nature, which causes water to form a spherical drop in order to touch as little surface as possible. Water spreads readily over hydrophilic (water-liking) surfaces. Most materials have hydrophilic surfaces when clean. The most common hydrophobic (water-disliking) surfaces are hydrocarbons, including oils and waxes, fluorocarbons, and silicones.

Regardless of the quality of the adhesive, its performance is dependent on the surface preparation and the design of the adhesive joint. In particular, clean surfaces are necessary to get the best results from an adhesive.

6.19.3 Fluidity of liquid adhesives

Rheology is the study of the flow of liquids. Thixotropy is the property of liquids to become temporarily more fluid when stirred due to the alignment of their tiny crystals. (An example is a suspension of clay in water.) The fluid thickens as soon as the stirring stops, and the crystals then return to their random orientation. This property of adhesives and paints helps to prevent a sag or run when these coatings are applied to vertical surfaces. Some polymers, such as polyamides, provide this property because of weak hydrogen bonds holding the polymer chains together.

The opposite of thixotropy is dilatancy. If the viscosity of a liquid increases with the shear stress of stirring, it is called dilatancy. This property is basic to transmission fluids used in the automatic transmission of automobiles. Both the viscosity and the surface tension decrease with an increase in temperature, so when possible, an adhesive is applied while hot so that it will wet and flow more efficiently.

6.19.4 Wettability

Wetting is another important property of an adhesive for it must wet the surface thoroughly as it spreads. The adhesive must be able to flow into the miniature surface crevices, displacing dirt, moisture, and trapped air, in order to have mechanical as well as chemical hesion. For better wetting of the surface, the surface tension of the adhesive can be lowered by adding a surfactant wetting agent.

6.20 Types of Adhesives

Rubber latex emulsions, made in the synthesis of chloroprene, styrene, and acrylonitrile rubbers, are used to make latex adhesives. Catalysts and vulcanizing agents are also needed to complete the formulation of rubber. Phenol formaldehyde, urea, and melamine formaldehyde resins are used extensively in adhesives. The epoxy two-package mix and the polyurethanes are becoming popular.

Four types of adhesives in general use for many years are organic solvent-thinned adhesives, latex adhesives (water emulsions), water-dispersed adhesives, and the two-package epoxy adhesives. Although organic solvents can be used to dilute these rubber-containing adhesives to get a thin film of adhesive, organic solvents can be a threat to our health and the environment. For these reasons, and for economy, the two-package type and the water-dispersed adhesives are more commonly used in industry.

6.20.1 Organic solvent-thinned adhesives

A contact adhesive is a good example of a solvated adhesive. The solvent cannot escape once the two nonporous materials are sandwiched together. Therefore, the solvent must be allowed to evaporate before the adherends are assembled. These organic solvent-thinned adhesives are applied to each of the two surfaces and the solvent is allowed to escape before the two surfaces are put together.

Solvents can be classified as active and nonpolar. An active, or good, solvent such as ethyl acetate or methyl ethyl ketone attaches to the polar groups of the polymer chains, thus breaking up the clumps of polymer chains. This allows the adhesive to wet the surface and form strong bonds. By contrast, a poor or nonpolar solvent, like an aliphatic paint thinner, only fills the pockets between clumps of polymer chains. These polymer clumps flow easily over one another as a low-viscosity fluid mixture with poor surface wetting or bonding. Thus an active solvent must be present in the formulation of latex adhesives to obtain adequate hesion.

6.20.2 Latex adhesives (water emulsions)

Latex film formers usually consist of natural or synthetic rubber or vinyl copolymers. These water-dispersed or latex adhesives contain, in addition to the elastomeric film former, many additives, including emulsifiers, thickeners, fillers, tackifiers, and antifoaming agents. Like paints, adhesives are proprietary systems and are carefully formulated. Synthetic rubber-based adhesives include styrene butadiene and neoprene. Neoprene is a specialty rubber with greater strength and solvent resistance.

6.20.3 Water-dispersed adhesives

Adhesives that are water-dispersed depend on natural materials for bonding, such as animal and vegetable glues. They remain readily dispersible in water after drying and the bond can easily be destroyed by soaking in water. Their use in industry is limited.

Although many of the natural adhesives have been partially replaced by synthetics, those made from animal and vegetable tissues are still in use. These include commercial glue made by the hydrolysis of collagen that is extracted from skins and bones of animals and fish. Their use is confined to porous materials. Natural adhesives, such as casein and soybean, are slightly more water-resistant and are used chiefly in the woodworking industry. Dextrin adhesives are made by the hydrolysis of starch for use with paper products. Rubber, whether natural or synthetic, is sometimes an additive in synthetic adhesives.

6.20.4 Two-package adhesives

With a limited shelf life, epoxy adhesives are made by using a low-molecular-weight partially polymerized epoxy polymer. By using the partially polymerized polymer, a smaller proportion of the skin-irritating amine is required to complete the polymerization process. The two-part package can be formulated so that an equal volume of the curing agent is used to form the adhesive. The exothermic reaction results in an increase in temperature and a slight decrease in the viscosity of the mixture. The viscosity is low enough to allow the adhesive to flow over the surface.

The disadvantage of a shelf life of only a few hours is more than offset by the strong bonds formed by the epoxy resin. Other advantages of the adhesive include the absence of a solvent to be evaporated and more water resistance. Epoxy adhesives are especially suited for use in the construction industry.

6.21 Advances in Structural Adhesives

Adhesives used in the construction industry must have high modular and tensile strength. The development of high load-bearing adhesives has led to their use in engineering applications. A familiarity with the chemistry, the structure, and the resulting properties of the resins used is important for selecting the appropriate adhesive.

The development of instruments to investigate the structure of composites, such as rubber-modified adhesives, glass ceramics, and particulate-reinforced plastics, has led to a major breakthrough in these fields. The use of adhesives for plywood and gypsum wallboard is commonplace. Adhesives with improved properties are used in industry to cement together ceramic, plastic, and metal objects (Figs. 6.7 to 6.9).

Structural adhesives are based on specially cured rubber-toughened epoxies, the acrylics, and the silanes. Rigid fibers and other particulates are used to achieve additional strength. Silane resins are used to prevent moisture penetration.

6.21.1 Rubber-modified resins

Epoxy adhesives, put into use after World War II, have grown rapidly in popularity. The epoxy monomer can be cured with organic amines, acid amides, acids, alcohols, or mercaptans to form epoxide resins. There are many varieties because the amine-curing agent can occur in so many forms. To avoid the brittleness of the usual epoxy resin, it must be modified to be suitable for use as a structural adhesive.

Improvement of resistance to fracture is obtained by the addition of reactive liquid rubber during the curing process. The cured rubber-modified resin has finely dispersed rubber phases, or regions, in an epoxy matrix. This might be likened to the formation of a grain structure in an alloy. Improvement in shear deformation and reduction in void formation and crazing are some of the results. Further improvement in adhesive strength is obtained by carefully controlling the temperature during the curing of the resin.

6.21.2 Acrylic adhesives

Solvent-free acrylic adhesives are made by the free-radical process. Other members of the acrylic family used in adhesives are cyanoacrylates and methyl methacrylate. They can be used as partially polymerized liquids with good shelf life and bonding properties and have a fast curing time.

Cyanoacrylates can be cured in place by the surface moisture on the substrate, except that metal surfaces with an oxide film can be a source of trouble.

Because these modified epoxy resins are not cross-linked, the adhesives made from them can be attacked by organic solvents and soften

Figure 6.7 Workers install tape for ceiling mirrors. Adhesive foam tape saves the time and cost of rivets, screws, or other mechanical fastening. (*Courtesy of 3M.*)

Figure 6.8 Adhesive tape holds up 9600 pounds of mirrored panels to the ceiling of an international airport terminal. (*Courtesy of 3M.*)

Figure 6.9 High-strength bonding tape holds aluminum wall panels to the aluminum frame, eliminating all visible mechanical fasteners on this house. (*Courtesy of 3M.*)

above 80°C (176°F). The addition of 5 percent acetic acid has been reported to increase their strength by 50 percent. Urethane methacrylates can be made as either hard or tough resins which, when mixed, make hard and tough adhesives. Cyanoacrylates, rubber-based acrylates, and modified acrylate resins make tough moisture-resistant fast-curing adhesives with good adherence to metal and most other materials. With research and development continuing, they have great promise for specialized uses.

6.21.3 Particulate-reinforced resins

The technology and understanding of the incorporation of rigid particulates in the matrix of polymeric resins have led to many improvements. Resistance to elastic and plastic deformation, fracture, and crack propagation, and toughness are all improved by the introduction of glass particulates into the rubber-modified acrylic and epoxy resin adhesives. Much research is continuing.

6.21.4 Silane surface primer

Extensive research has shown that silanes hydrolyze on metal and silica surfaces and then condense to form polysiloxane networks over the surface. It is believed that silane forms strong covalent bonds with these surfaces. Also, reactions of silane functional groups with epoxide and other resins have occurred. Consequently, silanes can be used as primers for a variety of surfaces. This surface preparation is followed with a coat of modified epoxide or acrylate adhesives to produce improved surface bond strength and superior wet durability for the adhesive joint. The development of special techniques is required for the application of this technology to structural adhesive joints.

6.22 Adhesives for Composites

Industrial adhesives need cohesive strength equal to, or greater than, the materials that are glued together. Since the adhesive is enclosed, the adhesive cannot solidify by solvent evaporation after the joint is formed. The solvent and thinner must either evaporate before the surfaces are sandwiched together or they will have to escape slowly through the surfaces. (This slow evaporation is possible only if the surfaces are of wood.) Phenol formaldehyde-like resins that cure without fluid by-products, or do not need to be thinned with solvents, are used to make composites.

6.22.1 Cross-linked adhesives

Phenol formaldehyde (PF) and urea formaldehyde (UF) adhesives are used for all but a small percentage of the plywood manufactured in the United States. Phenol formaldehyde is best for exterior exposures and urea formaldehyde for interior use. Because of the cost there are situations where the less expensive casein or soya glue can be used, especially in hot, dry climates. Unless they are treated with a preservative, they are subject to fungal attack.

6.22.2 Novalak resins

Novalak resins are phenol formaldehyde resins made under acid conditions. They melt and flow easily because they are primarily linear in structure. Resols are phenol formaldehyde resins made under alkaline conditions. When heated they are converted to cross-linked polymers that are infusible and do not swell in organic solvents.

6.22.3 The reesitol, and resital stages

Phenol formaldehyde thermosetting resins have been divided into three stages by international agreement. The alkaline-cured phenol formaldehyde resin in the A, or undercured, resol stage has a molecular weight of only a few hundred and is soluble in polar solvents. On further heating it forms the B, or intermediate, resitol stage. Due to branching and partially complete cross-linking, it is insoluble but swells in polar organic solvents. It is rubbery when heated. On further heating the resin is converted to the final resital, or C, stage. This is fully polymerized and completely insoluble, and no longer swells in solvents.

6.23 Health Hazards

Before using any product, ask for the Material Safety Data Sheet (MSDS) and read it. You should know the chemicals you are using and what precautions to take to prevent health and environmental problems connected with the product.

Sulfuryl fluoride, used in termite fumigation, is toxic by inhalation with a very low threshold limit value (TLV).

Most polymers are made from toxic monomers. When polymers are heated, especially linear polymers, they decompose to release their toxic monomers. If polymeric materials are disposed of by incineration, the noxious gases are lung and eye irritants and are toxic if inhaled over extended periods of time.

Polyurethane foam burns readily to produce hydrogen cyanide (a toxic gas) along with a dense, black smoke. Some polymers, such as phenol formaldehyde resins used in wallboards and carpets, are suspected of releasing formaldehyde by slow decomposition at room temperature. Over an extended period of time, formaldehyde can be a serious health hazard. When considering toxic substances, the length of time exposed as well as the concentration of the gas are both of vital importance.

Urea, melamine, and phenol formaldehyde resins are much used in industrial adhesives. Their slow decomposition at ambient temperature releases formaldehyde gas into the air. Formaldehyde, toxic by inhalation with a 1-part-per-million threshold limit value, is an irritant and a carcinogen. Epoxy resins are much used in the building industry because of their outstanding bond strength and durability. Epoxy adhesives have such a short shelf life (must be used a few minutes after mixing the ingredients) that most of them are marketed in two parts and are mixed just before using. Vapors in the uncured state are a strong irritant and cause severe dermatitis.

Most of the solvents are highly flammable and have a low flash point (ignition temperature). Many of the solvents that were previously used in adhesives are now banned by the EPA. These include chlorinated hydrocarbons, which are toxic to the liver, and the aromatic solvents, such as benzene and toluene, which are carcinogens. This has put pressure on the chemical industry to find replacement solvents. Instead of waiting until after one of the new solvents is shown to be toxic, be sure to take precautions to avoid unnecessary exposure, either in the workplace or in the home.

Some plasticizers slowly exude from polymers, evaporate from the surface of the plastic, or dissolve in detergents or other cleaning agents. The product eventually becomes brittle again. In addition, these plasticizers are toxic aromatic substances and usually carcinogenic. Long-time exposure to low concentrations of toxic vapors and carcinogens is of increasing concern to the public.

A good rule to remember: if you can smell a chemical, you should avoid exposure to it.

6.24 Recycling of Wood, Paper, and Plastics

Instead of sending used wood to the dump, contractors and builders can either reuse the wood for other purposes or reduce it to chips for garden compost.

Paper, including newspapers, white office paper, magazines, and cardboard, can be recycled successfully. Some communities and offices are successful at recycling; others are in the "thinking about it" stage. The average office worker discards over 100 pounds of paper each year, which amounts to an estimated 30 million trees. Much of that paper could have been reused as scratch paper and then sent on to the recycling center. Businesses can make a small profit from the recycling of white office paper. Approximately 7 tons of old newsprint was recycled in 1991 in the United States. (Most newsprint went to landfills.) We must realize that we are not doing enough to reduce our waste stream.

Communities faced with limited space in landfills and increased concerns about the disposal of hazardous materials are pressuring industry to reconsider the basic purposes of performance and profitability. Now there is concern about the products after they become obsolete and are discarded. Some communities have gone so far as to threaten to ban the sale or use of materials which cannot be recycled. An example is the attention focused on polystyrene (PS), a popular packaging material. The plastics industry pointed out that polystyrene makes up less than 0.3 percent of the municipal waste.

TABLE 6.2 Plastics Recycling Information

Number	Symbol	Name	Common use
1	PET	Polyethylene terephthalate	Soft-drink bottles, microwave food bags and trays, packing film
2	HDPE	High-density polyethylene	Trash bags, milk cartons, soap and bleach bottles, pipes and molded fittings
3	V	Poly (vinyl chloride)	Plastic wrap for meats, cooking oil bottles, conduits, plumbing pipes, siding, gutters, window frames
4	LDPE	Low-density polyethylene	Grocery store vegetable and food wrap, wire and cable coatings, insulation
5	PP	Polypropylene	Packaging film, housewares, auto parts, air filters
6	PS	Polystyrene	Foamed packaging and insulation, refrigerator doors, air conditioner cases, radio and TV cabinets
7	Other	All other polymers	Various

However, polystyrene industries responded by promoting several major polystyrene-recycling industries. To their goals of production and profit, they have added a concern about what happens to their products when they are worn and obsolete.

The recycling of discarded materials requires several essential steps. Discarded plastics must be collected, before or after sorting, into a uniform feedstock for recycling. Then they must be transported to the nearest recycling plant for treatment. Finally, the treated raw material must go to a manufacturing plant, which faces stringent EPA and local code restrictions. Equally important, they must have ready markets for the recycled material to go into useful objects acceptable to the public. Each step along the way poses technological problems. The building industry is facing these same problems with much to be considered, for example, the challenges of demolition, recycling, and rebuilding of slum and deserted areas.

A commendable step in the sorting of plastics has been the general acceptance of a code for plastics manufacturers. The code is embossed on each plastic article sold. Two groups have pioneered this system of labeling plastics: the Society of Plastics Industry and the ASTM (D 1971). Table 6.2 provides a summary of the common types of plastics, their identification numbers, and their symbols.

Terms Introduced

Adhesion	Isoprene
Adhesive primers	Lignin
Butadiene	Particleboard
Cellulose	Polybutadiene
Chloroprene	Polymer transition temperature
Cohesion	Polymers
Dilatancy	Rheology
Elastomers	Surface tension
Fungi	Termite control
Hesion	Thixotropy
Hydrophilic surfaces	Wood composites
Hydrophobic surfaces	Wood rot

Questions and Problems

6.1. Distinguish between adhesive and cohesive bond failure.

6.2. List the typical ingredients in an adhesive.

6.3. What properties of the adhesive and the substrate are necessary for good hesion?

6.4. What is meant by solvated, emulsified, and two-package adhesives?

6.5. Distinguish between polar and nonpolar substances and give two illustrations of each type.

6.6. Distinguish between isotactic, syntactic, and atactic polymers.

6.7. Explain what is meant by the contact angle of water on a surface and describe one of its uses.

6.8. Distinguish between thixotropy and dilatancy.

6.9. Why is the expansion coefficient of the substrate important in the selection of an adhesive?

6.10. How is it possible to reuse rubber in retread tires?

Suggestions for Further Reading

ASCE, *Structural Plastics Selection Manual,* ASCE Manuals and Reports on Engineering Practice no. 66, American Society of Civil Engineers, New York, 1985.

Ronald S. Bauer, *Epoxy Resin Chemistry II,* American Chemical Society, Washington, D.C., 1983.

L. Reed Brantley, "Removal of Organic Coatings," *Industrial and Engineering Chemistry,* vol. 53, p. 310, 1961.

L. Reed Brantley, "Factors Affecting Adherometer Adhesion," *Journal of Chemical and Engineering Data,* vol. 6, p. 288, 1961.

A. J. Kinloch, ed., *Structural Adhesives,* Elsevier, London, 1986.

G. C. Mays and A. R. Hutchinson, *Adhesives in Civil Engineering,* Cambridge University Press, Cambridge, 1992.

J. A. Panek and John P. Cook, *Construction Sealants and Adhesives,* 3d ed., John Wiley & Sons, New York, 1991.

Gerald L. Schneberger, ed., *Adhesives in Manufacturing,* Marcel Dekker, New York, 1983.

Terry Seller, *Plywood and Adhesives Technology,* Marcel Dekker, New York, 1985.

Raymond Setmour and Charles Carraher, *Structure—Property Relationships in Polymers,* Plenum, New York, 1984.

W. Wayne Wilcox, Elmer E. Botsai, and Hans Kubler, *Wood as a Building Material,* John Wiley & Sons, New York, 1991.

Robert A. Zabel and Jeffrey J. Morrell, *Wood Microbiology: Decay and Its Prevention,* Academic Press, San Diego, Calif., 1992.

Chapter

7

Roofing, Sealants, and Fire Protection

7.1 Introduction	218
7.2 Roofing	218
7.3 Understanding Emulsions	221
7.3.1 Example of Formation of a Colloidal Solution	223
7.3.2 Example of Formation of an O/W Emulsion	223
7.3.3 Example of Cracking an Emulsion	224
7.4 Membrane Materials	224
7.4.1 Asphalt Derivatives	224
7.4.2 Asphalt Emulsions	225
7.4.3 Asphaltic Bitumens	225
7.4.4 Blown Asphalt	225
7.4.5 Cracked Asphalt	225
7.4.6 Cut-Back Asphalt	225
7.5 Built-Up Membranes	226
7.5.1 Single-Ply Membranes	226
7.5.2 Foamed Plastic Membranes	227
7.6 Water-Shedding Systems	227
7.7 Residential Roofs	228
7.7.1 Shakes and Shingle Roofs	228
7.7.2 Asphalt Shingles	229
7.7.3 Grades of Asphalt Roofing Materials	229
7.7.4 Wood Shingles	229
7.7.5 Mineral-Fiber Shingles	230
7.8 Insulation of Roofs	230
7.9 Typical Problems and Correction	231
7.10 Government Restrictions	233
7.11 Recycling Roofing Materials	234
7.12 Warranties	234
7.13 Sealants	235
7.13.1 Movement Capability	236
7.13.2 Hardness	237
7.13.3 Adhesive Promoters	237
7.13.4 Primers	237
7.14 Sealant Types	238
7.14.1 Oil-Based Caulks	238

7.14.2 Polysulfide Sealants	239
7.14.3 Butyl Sealants	239
7.14.4 Acrylic Sealants	240
7.14.5 Urethane Sealants	241
7.14.6 Silicone Sealants	242
7.15 Fire Protection	243
7.15.1 Materials at High Temperatures	243
7.15.2 Wood Exposed to Fire	244
7.15.3 Concrete Exposed to Fire	245
7.15.4 Steel Exposed to Fire	245
7.15.5 Polysteel Forms	246
7.16 Fire Security	246
7.16.1 Fire Detection	247
7.16.2 Containment of Fire	247
7.16.3 Protection of Structural Elements	248
7.17 Repair of Damage	248
7.18 Health Hazards	250
Terms Introduced	251
Questions and Problems	251
Suggestions for Further Reading	251

7.1 Introduction

The materials and operations introduced in this chapter will add very little to the cost of a building, but they are essential for its satisfactory performance. Many of the common problems of roofs and their corrections are investigated. The experiments with emulsions will help explain the nature of the chemicals being used. We detail the reuse, disposal, and recycling of used roofing materials.

Sealants are compared and tips are given on methods of preventing costly water destruction of the interior walls and furnishings. The importance of moisture and air-draft barriers, representing a small fraction of the cost of construction, is discussed. Furthermore, advice is given about how a meticulous use of sealants can save on the cost of energy used for heating and cooling.

The chapter concludes with the principles and methods of fire protection, the fire ratings and resistance of building materials, fire detection, containment of a fire, repairs of fire damage, and health hazards.

7.2 Roofing

The purpose of a roof, representing only a small fraction of the total cost of a structure, is to protect the building and its occupants from the harmful effects of the environment. In their simplest form, roofs can be classified either as built-up membrane systems or as water-shedding systems. A built-up membrane system might have a top layer of roofing gravel, a built-up membrane, a thermal insulating layer, and a polypropylene sheet as a moisture barrier next to the roof deck. Water-shedding systems designed to keep out environmental elements include thatch, shingles, slate, tile, metal, plastics, and gold. Some are inexpensive, some are extremely expensive, and many have been used successfully for centuries.

Homeowners and owners of commercial buildings expect a roof to give many years of satisfactory trouble-free service. However, the failure rate of roofing systems is estimated to be half of that of all installations. This has encouraged the development of new systems. And of the hundreds of new roofing systems and materials promoted over the past few years, most are no longer used.

Roofing materials range from the traditional bitumen-asphalt-felt covering to innovative plastic membranes. The most popular roofing materials in use today are built-up, single-ply, and modified bitumen, but many other kinds of roof coverings are used, namely, metal sheeting, corrugated metals, plastic composites, ceramic tile, wood, and shingles of all sorts.

An unconventional roofing system is solar roofing, a specialty of its own. The two techniques used are burying the building and flooding the roof. Earth-covered homes are seen in most states in the United States and can be practical. Construction costs are high due to the structural support required to hold tons of earth on top and at the sides. Heating and cooling costs are low since the earth covering keeps the indoor temperature at around 20°C (68°F) all year.

Flooded roofs are seen in warm-weather states where there is little chance of freezing. The building's roof, lined with a plastic membrane, is flooded to a depth of 61 centimeters (24 inches) with a water-salt or water-antifreeze mixture. Heat from the sun is collected throughout the day. A motor-driven cover rolls over it during the evening, rolling back off at dawn. Cool water circulates through the building walls during the day and warm water circulates during the night. Added structural support and routine maintenance (filtration, care of algae, and added equipment cost) are disadvantages.

Plastic, used for roofing, has very strict building codes. Plastic burns and when melting, it aids in the spread of fire. Areas not considered living areas, such as patios, sun porches, swimming pools, carports, and barns, may have plastic roofing subject to code approval.

Slate and stone shingles, on many ancient structures in Europe, may last for centuries.

Figure 7.1 Thatched roofs are popular and well maintained in rural Ireland. Painted walls are solid stone from the area.

Thatch, still used in some parts of the world, has been popular for centuries (Fig. 7.1). The inexpensive materials used are straw, rushes, reeds, or certain kinds of palm leaves. This material is combustible. Local codes may prohibit its use.

At the other extreme, gold foil or gold leaf is a feasible covering for very expensive homes and government buildings. Although expensive in bulk, gold is so malleable and resistant that it can be used in sheets only a few mils thick (a mil is one thousandth of an inch) and pressed to fit snugly over steel or other metal surfaces. One of the chemists' "noble" or inert metals, gold does not corrode or change from its spectacular color. Although there could be some question about its resistance to abrasion, gold foil is said to be easy to maintain and theft free. (The thin foil clings so tightly that it is useless to try to remove it.)

Sheet metal roofs are of industrial importance. Their advantages are strength, fire resistance, resistance to mold and insect destruction, environmental resistance, and thus resistance to water leaks. Sheet metal roofing material usually comes with built-in joining lips or seams. Their disadvantages include the chemical reaction of the tannic acid in redwood and cedar support beams and the fact that ferrous metals rust and must be protected by galvanizing with zinc. Copper and aluminum sheet metal do not rust. Copper reacts with the air to form a green coating thought by many to be decorative. Since metal has a larger expansion coefficient than wood or concrete, but much less than plastics, expansion joints may be needed for large roofs.

Other disadvantages of metal or corrugated plastic roofs are that they readily transmit vibrations through the sheathing and supporting beams. They act as sounding boards to transmit disturbing noises to the occupants during heavy winds, rain, and hail storms. Prevention involves securing the ribs to the supporting sheath with plastic gaskets or foam rubber strips. This added anchorage against flexure and vibration also increases the life of the corrugated roof.

In spite of the wide choice of durable materials to be used, failures are frequent, serious, and more expensive than most design professionals suspect. The following statistics indicate the widespread problem.

1. Facade failures have increased to more than 33 percent of all insurance claims.
2. Nearly 65 percent of all high-rise condominiums have experienced major parking garage problems—water leakage and concrete deterioration.
3. 25 percent of high-rise condominiums have roofing problems; 40 percent of townhouse roofs need repairs.
4. More than half of all high-rise condominiums and townhouses experience water leaking through the walls.
5. Costs of repairs may vary from $5000 to over $200,000.

Because of these problems, many owners have gone to court to recover costs from the developer or contractor. A little more time and money spent for additional pages of details could save more money by avoiding time-consuming and expensive litigation proceedings. The proper design of a roof, the right materials, and knowledgeable workmanship are extremely important if expensive problems are to be avoided.

7.3 Understanding Emulsions

Asphalt, an emulsion, is a product used extensively in the roofing industry. In order to understand the behavior of asphalt, we must first understand emulsions.

"Oil and water do not mix." This is an old saying, but oil and water are easy for chemists to mix. They use a small amount of an emulsifier to disperse (separate into fine droplets) one immiscible liquid inside another liquid.

A detergent (sodium lauryl sulfate) is a good emulsifying agent for making an emulsion of asphalt in water. The detergent makes the outside of the asphalt droplets hydrophilic (waterlike). The finely dispersed droplets of asphalt make up what is called the internal phase (internal liquid) and water is the continuous external phase of this emulsion. With their waterlike coatings, the asphalt droplets now mix readily with water to form an emulsion.

An oil-in-water emulsion is abbreviated O/W. W/O is the abbreviation for a water-in-oil emulsion.

The purpose of an emulsion is to spread a semisolid liquid, such as asphalt, more easily. For example, a stiff asphalt is difficult to spread unless it is converted into an O/W emulsion. In general, the viscosity and other physical properties of an emulsion become those of the external phase. The asphalt emulsion can then be spread easily, like water.

Emulsions are limited only by the number of separate phases (physically distinct solids, liquids, or gases) which are available to disperse within one another. A phase must be homogeneous (uniform to the eye). We make O/W emulsions each time we wash (emulsify) the oil and dirt from our skin with a soap or detergent.

Egg yolk (containing lecithin) is a household emulsifier. The lecithin in the yolk makes finely divided droplets of water hydrophobic (oil-loving) by causing the outside of the droplets to be oily. This oily surface causes the water droplets to mix easily with oil; one example is mayonnaise.

To generalize, emulsions are but one of a class of colloidal systems in which one phase is dispersed into another. To help stabilize the dispersed particles, the particles are covered with an emulsifier, a peptizing agent, or an electric charge if the particles are suspended in air. These coatings help keep the dispersed particles more compatible with the external phase and keep them from coagulating.

A colloid, such as an emulsion, looks like a new material, but otherwise has many of the physical properties of the external phase. Besides asphalt roofing emulsions, composites such as clay, grout, steel composites, and wallboard are examples of colloidal dispersions of one phase in another, as shown in Table 7.1. The use of other colloidal systems is discussed in other chapters.

The viscosity of emulsions is that of the external phase, but the color depends on the fineness of dispersion (like the color given a sunset by fine dust dispersed in the sky). An emulsion becomes whiter as the droplets become smaller because of an increase in the mirrorlike reflections of light. In general, colloidal systems do not resemble either of the phases they were made from—they appear to be new substances.

TABLE 7.1 Classes of Colloidal Systems

Class	Example	Internal phase	External phase
Aerosol	Smoke	Solid	Air
	Fog	Water	Air
Sol	Shellac	Solid	Liquid
Foam	Bubbles	Air (gas)	Liquid
	Styrofoam	Air	Polystyrene
Emulsion	W/O	Water	Oil
	O/W	Oil	Water
Composite	Plastic	Fibers	Polymer
	Metal	Fibers	Metal

7.3.1 Example of formation of a colloidal solution

Directions. Fill two 1-liter beakers (or other Pyrex jars) three-quarters full of water. Heat the water in one beaker to boiling and stir while adding a drop of saturated ferric chloride solution. Do the same to the cold water in the other beaker. Compare the colors in the two beakers. Add another drop or two of ferric chloride solution to each beaker and observe the color changes. Continue adding ferric chloride until the red color no longer deepens in the hot water. Adding a drop of a gelatin solution will act as a protective colloid (peptizer) and will stabilize the colloid (sol) with a protective coating of gelatin.

Interpretation. To make the colloidal solution (sol), the water is heated to boiling to hasten the hydrolysis of the ferric ion from the ferric chloride to form ferric hydroxide and hydrogen ion. The precipitated ferric hydroxide molecules coagulate to form particles that scatter light in the visible red frequency to make the sol appear red.

Comparing the color developed in the boiling water with that in the cold water shows that the color of the sol is not the yellow ferric ion; this depends on the particle size. By shining the beam of a flashlight through the solution, the path of the light is seen the same way as a beam from a searchlight appears in the night sky.

7.3.2 Example of formation of an O/W emulsion

Directions. Add a few drops of a household detergent to the water in a jar and shake vigorously to form a foam. Then add a few drops of mineral oil and, again, shake vigorously. Add detergent until you see a milky emulsion.

Interpretation. The household detergent (sodium lauryl sulfate) is a cationic (negatively charged) emulsifying agent and is adsorbed on the oil droplets, making them hydrophilic. The result is an O/W emulsion.

7.3.3 Example of cracking an emulsion

Directions. Pour half of the O/W emulsion into another jar and, while stirring, add a few drops of hydrochloric acid until the white emulsion separates to form oil drops. An alternative is to crack (coagulate) a mixture of half milk and half water with a little vinegar. (*Note:* The oil, milk fat, separates from the water on standing.) In order to crack an emulsion, the emulsifying agent must be altered by freezing or by an acid reacting with the emulsifier.

How can you tell whether an emulsion is O/W or W/O? Spread a little butter over the palm of your hand. Add a drop of water to the area. If the water remains as a spherical drop, the butter is W/O. If the drop of water disappears into the film of butter, it is O/W. Try this test on shaving cream or face cream.

Interpretation. Milk is an O/W emulsion. As the bacteria in milk cause the milk to sour (by forming lactic acid from the milk sugar), the homogenized milk emulsion also cracks. Freezing or adding salt is another way to crack this emulsion.

7.4 Membrane Materials

Asphalt is a dark-brown to black emulsion. It is a semisolid or solid cementitious waterproofing material. Asphalt contains high-molecular-weight aliphatic, aromatic, and heterocyclic compounds. The main components are bitumens. Besides petroleum asphalt, there are bituminous coal asphalts and Trinidad asphalts.

Native asphalt crude is converted to residual asphaltic bitumen by distilling off the more volatile naphtha, gasoline, and aromatic fractions. The name asphaltic bitumen is a combination of the terms asphalt (formerly used in the United States) and bitumen (used in England).

7.4.1 Asphalt derivatives

There are two kinds of asphalt: naturally occurring asphalt and an asphalt residue which is obtained from the distillation of California petroleum. Common modifications are asphalt emulsions, asphaltic bitumens, blown asphalt, cracked asphalt, and cut-back asphalt.

7.4.2 Asphalt emulsions

Asphalts, or bitumens, are readily dispersed in water to form an O/W emulsion. The dispersing agent is often the common sodium lauryl sulfate detergent with proprietary modifiers. Asphalt is a mixture of aliphatics, aromatics, and ring compounds containing sulfur, nitrogen, and oxygen and comes from petroleum residues, in contrast to the tars, which come from the carbonization of coal and wood. Two types of bitumen, asphalt and coal tar, are used in roofing and asphalt paving. The aromatic compounds are carcinogens.

7.4.3 Asphaltic bitumens

Asphaltic bitumens are prepared from pitch or unrefined petroleum. This is a colloidal suspension of high-molecular-weight hydrocarbons dispersed in their much lower-molecular-weight constituents. Note that the external phase determines the viscosity of an emulsion. In a sense, this product constitutes the remnants left after everything else of commercial value has been skimmed off. This residue is mainly used for roadbeds.

7.4.4 Blown asphalt

As the name implies, this blown (oxidized) asphalt is obtained by blowing air through heated asphaltic bitumen. This is the refined residue when the sulfur compounds and unsaturated compounds are oxidized by the oxygen of the air blown through it at 250°C (482°F). In the process the more volatile sulfur dioxide and organic acids are removed.

7.4.5 Cracked asphalt

Cracked asphalt is prepared by heating the asphalt to about 450°C (842°F). This product has most of its carcinogenic components destroyed or removed. It is used to prepare a tarred roofing membrane. Over the years, workers and children have been known to chew cracked asphalt (an unhealthy practice, of course).

7.4.6 Cut-back asphalt

Asphalt that is diluted (thinned) with hydrocarbon solvents is known as cut-back asphalt. With its low viscosity it seeps into cracks and pores of the concrete. It is easy to spread and is used to waterproof and cement, or bond, a membrane felt to a concrete deck or flashing. The hydrocarbon solvent-type thinner can be a naphtha, gasoline, or jet fuel, depending on availability and cost. In many instances these organic thinners are now replaced by water-thinned dispersions (emulsions) because of strict EPA and environmental concerns.

7.5 Built-Up Membranes

Built-up custom-built roofing consists of a structural deck, thermal insulation, and a roofing membrane. A number of alternating layers of bitumen and felt make up the weather-resisting membrane. There is a vast market for the skills of qualified contractors to apply built-up roofing on residences and commercial properties (Fig. 7.2).

7.5.1 Single-ply membranes

The simplicity of single-ply membranes accounts for their popularity. One layer of a watertight membrane is sealed at the seams and edges. The single-ply sheeting can be made from several materials, such as rubber or plastic. Modified bitumens are composed of poly-

Figure 7.2 Built-up roof having flat pitch and gravel roof membrane and showing vent pipe, parapet wall, and flashing overlaps with sealant (a common source of water leaks).

ester fibers or fiberglass felt, which comes in a roll. Bitumen provides the waterproof binder.

7.5.2 Foamed plastic membranes

Various foamed polymers, such as polyurethane, are in use. Although these are crushed easily, foamed plastic membranes provide superior thermal insulation. Foot traffic should be kept to a minimum unless protection can be provided by walkways. Adhesives to hold the walkway pads or coverings in place are a terpolymer elastomer made from ethylene-propylene diene monomer (EPDM) or a poly(vinyl chloride) roof adhesive.

7.6 Water-Shedding Systems

Many complaints are caused by insufficient attention to waterproofing. Some of the common design and construction problems are insufficient slope to drain water from the roof, leaky flashing around the parapets or around vent pipes, inadequate waterproofing around doors, elevator shafts of tall buildings with inadequate flashing, or machinery on the roof not securely anchored in place. This is in addition to problems of the load-bearing capacity of the roof membrane and improper choice of roofing materials.

Some slope in a roof is necessary for proper drainage. A flat roof must have a slight incline. Such slightly sloping roofs are built on dead-flat decks by using tapered blocks of foam insulation under the built-up roofing. Water can flow to drains at either the sides of the roof, the center of the roof, or the edge of the elevator shaft, or specially designed drains can be located in the four corners of the roof. The inch-deep puddles that are found on roofs after a rain may be due to too small a slope to compensate for sag either resulting from deterioration of the foamed insulation, or occurring in the layers of the bitumen-treated membrane.

Other possible problems are that the metal flashing shield may not go high enough or far enough under the membrane, or the caulking could have cracked or pulled loose from the parapet wall or vent pipe. The difference in the thermal expansion coefficients of the flashing and the surface to be waterproofed requires the use of the highest-grade sealant that is available. If flashings are attached to wood, the nails could have pulled away to open a gap for the wind to drive in the rain. Not only is flashing needed at the edges of the membrane, but special care must be taken in the valleys formed when two sections of the deck form an angle to one another.

Fire escape doors to the roof should be recessed or sheltered from the weather to prevent them from rusting. The threshold should be high enough to avoid a difference in air pressure to keep water from being sucked into the stairwell. A strong automatic closure device is vital for weather and fire protection.

The sheathing used, that is, the part of the roof that supports the insulation and the roofing membrane, depends on the type of building. A good grade of planking or five-ply sheathing is recommended for homes. The underside of the sheathing may serve as the ceiling in some homes and will determine its finish and appearance. For commercial buildings, plywood, metal, or concrete is usually used. Concrete needs a sealer of solvent-thinned bitumen or a water emulsion of bitumen, and must be thoroughly dry before laying the felt covering.

Many flat roofs of concrete will have parapets (concrete walls) around the edge of the roof. The purpose is to hide vents and equipment on the roof as well as to provide some safety to people who must work on the roof. These walls must be strong, for they are often used by painters and window cleaners to anchor their equipment. Waterproofing this edge of the roofing membrane with flashing is a challenge because this area is frequently the source of water leaks into the floor below.

Vibrations from machinery, such as heat pumps, solar heaters, fans, and other equipment, cause leaks over the years, and the noise is objectionable to the occupants of the building. To protect the membrane, walkways should be provided for access to equipment for maintenance workers. Careful installation of flashing is essential where the equipment comes into contact with the roof membrane.

Note that an infrared camera can be used to find water leaks in a roofing membrane.

7.7 Residential Roofs

Roofs on residential buildings may be from a wide choice of materials. These include wooden shakes; shingles made of asphalt, fiberglass, or wood; tiles of cement, clay, or slate; and asphalt roll and gravel. Selection of the type depends on weather conditions, neighborhood roof styles, price, and style of home.

7.7.1 Shakes and shingle roofs

Shakes and shingles are limited to use on sloping roofs. Shakes are made from wood split along the grain and have a characteristic rough appearance. Shingles are sawed from wood, but can also be made from many other materials in order to be more fireproof, thereby lowering fire insurance rates.

7.7.2 Asphalt shingles

The fire and wind resistance; the wide variety of colors, texture, and design; economy; and an average 20-year life expectancy of asphalt shingles and rolls make them the most used roof-covering material in the United States. The Underwriters Laboratories use an A, B, C rating system for asphalt material. To receive their label of fire resistance, the material must pass flame-spread tests. Grades A, B, and C are each labeled fire-resistant. To be labeled wind-resistant, the roof or its shingles must resist being lifted or torn off by the prevailing or expected maximum winds of the area.

Asphalt shingles come in a variety of shapes, based on the way they fit together. The common varieties are the standard rectangular and hexagonal shapes, and these come with interlocking tabs.

7.7.3 Grades of asphalt roofing materials

Fire-retardant roof-covering materials can receive the UL label from Underwriters Laboratories if the manufacturer's product qualifies. Use of their label indicates satisfactory saturation, thickness of coating, and penetration of the asphalt; distribution of surface fire-protective granules; color, weight count, and size; and instructions for application. Asphalt shingles and sheet roofing should have the UL quality label since this is the recognized standard for the industry.

Class A is reserved for roofing materials that are effective under severe fire conditions. Under specified conditions, they are not readily flammable; they do not carry or transmit fire; they offer reasonable fire protection to the roof deck; they do not slip and expose the layer below; they provide no flying fire brands; and they do not require frequent repair.

Class B is reserved for roof coverings which are effective against moderate fire exposures. Under such exposures, they are to offer similar protection as class A.

Class C is used for roof coverings which are effective under light fire exposures. Under such exposures, these coverings are to afford similar protection as the other two classes.

7.7.4 Wood shingles

Red cedar is the preferred wood for shingles because of its small grain and absence of knots or other blemishes, the relatively little weight it adds to the roof, and the small amount of warping and cracking on exposure to the weather due to low moisture absorbency.

Cedar shingles are sold in three grades. The top grade, no. 1 blue label, is made from choice cedar and recommended for roofs and sidings that are fully exposed to the weather. No. 2, red label, is recommended for limited exposure to weather since it has slightly lower standards. No. 3, black label, is a utility grade for economy applications.

7.7.5 Mineral-fiber shingles

Four classes of mineral-fiber shingles conform to the requirements of ASTM C 222: individual unit, multiple unit, Dutch unit, and ranch unit. These classes differ from each other chiefly in use, shape, and size.

7.8 Insulation of Roofs

The purpose of insulation is to provide a water and heat barrier. The water barrier prevents rain leakage into the building. The heat barrier serves two purposes. It reduces the loss of heat from the interior of the building through the roof, and it prevents condensation of dewlike moisture on the ceiling beneath the roof. The moisture condensation occurs in freezing weather if the temperature of the ceiling falls below the dew point of the heated moist air in the building. Such moisture promotes the growth of fungi and mold, which weakens the wood supporting the roof structure.

Most roofs have insulation in direct contact with the roof decking or in between the ceiling joists. Many private homes have steep roofs, which provide attics for easy application of insulation.

Nearly half of the heat loss in a home is due to cold air from the outside coming through cracks and crevices. These must be caulked and insulated. Loose-fill insulation usually consists of rock wool, mineral wool, or fiberglass. The loose material is poured or blown into inaccessible areas such as areas in the attic, wall cavities, and hollow-block masonry walls. Rolls or mats of insulating material are laid between the rafters while avoiding filling the air vents, roof ventilators, and electric wiring.

When attics are used for storage or living space, the insulation is fixed to the roofing deck, covered with a built-up asphalt-impregnated felt membrane, which is protected with asphalt shingles or pebbles and gravel. Roofs with a good slope can be covered with overlapping wooden shingles, which are not watertight, but serve to guide the rain or snow off the roof.

Flat roofs should have the insulating material attached directly to the roofing deck. The insulation needs to be protected from water infiltration with the usual built-up roof covering. (Commercial buildings with the need to conserve space often have a flat roof, no attic, and insulation on top of the roof deck.)

With a comfortable living temperature on the top floor of the building and possible subzero temperatures on the outside of the roof, a temperature difference between the deck and the first membrane layer of 38°C (100°F) can cause thermal contraction of the top bitumen felt layer in addition to making the bitumen brittle and fragile. This threatens the bond holding the membrane layers together. In hot weather, a few drops of water filtering between the two top layers of the membrane can vaporize to form a bubble, threatening the bond between the layers.

7.9 Typical Problems and Correction

The following questions are asked frequently:

1. Why is water coming down the wall in a room on the top floor of a building? Flashing against the parapet wall on the roof has probably pulled loose and the opening allowed rain to run down the wall to the floor below. The flashing at parapet walls is probably the most common source of roof leaks, and regular maintenance could prevent this complaint. Metal flashings should not exceed 3 to 4 meters (10 to 12 feet) in length because thermal expansion can cause them to buckle and pull away from the parapet walls. To repair, anchor the flashing securely to the wall and caulk the edge touching the wall with a flexible caulking material. If the metal needs to be replaced, a flexible plastic flashing will be less apt to be a problem.

2. Why does the attic ceiling have patches of mold covering it, but no sign of water leaks? Evidently the temperature of the ceiling has been chilled below the dew point by the freezing temperature on the rooftop. The moisture condensed on the ceiling from the hot, humid air in the attic. The wet areas of the ceiling caused mold to form and thrive on the ceiling. Correction for this problem is to provide more or better insulation for the roof so that the extreme cold on the roof does not chill the ceiling below the dew point.

3. What is the cause of most roof leaks? Leakage at the parapet wall and the junctures of roofing materials with vent pipes and rooftop equipment are frequent causes of roof leaks. Regular maintenance could prevent this complaint. Drains should prevent water from standing in this area. Thermal expansion and contraction cause

the metal flashing to pull away from the parapet wall, allowing rain to run down the wall to the floor below. The solution is to press the flashing into the wall—sometimes a groove is provided—and recaulk.

4. What should be done about a large blister (bubble) on a built-up roof? These are spectacular, but usually do not mean that there is a membrane failure and water leakage. They are inflated by the expansion of water vapor. Once the top layer of felt is raised, it stiffens and remains inflated. The bubble can be pierced, gently deflated, and the hole where it was pierced can be sealed with a covering of asphalt emulsion. However, if there is little danger of the bubble being torn open, it can be left undisturbed since it is not a source of leakage.

5. What should be done if water is ponding around the drain on the edge of the roof, especially during heavy showers? Evidently a scupper (a drain going into the wall) was overlooked. Scuppers in a parapet wall behind and above the level of the surface drain are usually supplied as an emergency overflow in case the surface drain is clogged. The drain should be unclogged and a scupper constructed in the wall.

6. The ceiling is leaking near the center of the building, where the electrical and drainage facilities service the building. The pipe connected to the roof drain is clogged with roof gravel and the extra weight of the gravel has pulled apart a pipe connection. Gravel is put on top of a built-up roof as a protective sun screen and to help hold down the membrane in order to keep it from blowing away during heavy winds. A low curb near the edge of the roof can keep the gravel from being carried off the roof by heavy rains, ice, and snow. Roof drains should be covered with a coarse grating to keep gravel from filling the drain pipes. Gravel is becoming scarce and is being replaced by a highly reflective aluminum paint to reflect the sun's rays and avoid excessive heating.

7. What should be done about puddles of water standing on an asphalt built-up roofing membrane after a rain? To ensure drainage, roof membranes should have a gentle slope toward the drains. Too often, perhaps due to poor design or construction or to repeated roof coverings with one layer on top of the other, the layers of felt are lower in some regions and provide a depression for water to collect. Although these are usually tolerated, they can be filled with a few patching layers of roofing felt carefully cemented down with an asphalt emulsion or cut-back asphalt.

8. Water is leaking down the vent pipes that emerge from the roof. Lead sheet flashing has pulled away from the pipe or is not tightly wrapped around the pipe. Lead is malleable and can be readily shaped to the pipe.

9. There are ceiling water stains, but no leaks have been found around the edge of the roof or the drains. There is some danger of a leak developing wherever the roof changes slope. The flashing may be too small and difficult to caulk because of insufficient overlapping of materials.

10. Shingles (wood or asphalt) are ripped from the roof in a strong windstorm. The shingles may have been installed with too much of the shingle showing (insufficient overlap). Also, the shingles may have been stapled in place mechanically with a varying depth of penetration into the roof. A better practice in regions of frequent strong winds is to hand nail the shingles and individually tab or cement them into place. In hurricane-force winds roof coverings are likely to be ripped loose. In this case even the hurricane straps holding the roof structure to the wall beams might loosen and let portions of the deck blow away.

7.10 Government Restrictions

To meet the ever-increasing requirements and restrictions of OSHA and the EPA, many long-established practices are being challenged. The necessary changes in materials manufactured for roofing require new designs and technologies for installation and repairs. OSHA is concerned about the safety of workers using hot inflammable asphalt and about the inhalation of fumes.

Several of the common roofing materials, including asphalt, are suspected or proven carcinogens and others are being added to the list. Workers are provided with, and urged to read, the material safety data sheets (MSDSs) on all materials they use. It should be noted that the effect of exposure depends not only on the amount of chemicals in the air being breathed, but also on the length of exposure because the effects are cumulative. Also, the effect can be synergistic, that is, the effect of one chemical can be increased by the presence of another, different chemical.

The volatile organic compounds (VOCs) contained in some roofing materials are continually being reduced by new EPA regulations to reduce air pollution. For example, solvent restrictions on thinned cutback asphalt preparations require new design procedures for roofing materials to meet the Southern California requirements. As in the paint industry, one resolution is to replace solvent-based systems with water-based (emulsion) coatings. Other promising approaches are to combine synthetic fibers with adhesives to form roofing granules, which are formed into rolls of roofing sheets containing heat-sealing properties with little or no emissions of volatile organic compounds.

7.11 Recycling Roofing Materials

The proper disposal of used roofing materials is an increasing problem. Because of their slow disintegration in landfills, especially rubber-type materials, some landfill agencies require that used sheeting be cut into small pieces. With the number of landfills diminishing and the increasing concern about chemicals leaching into the water supply, more environmentally friendly, new synthetic materials are being chosen, resulting in a recycling problem with the discarded materials. Recycling facilities are too few, but asphalt shingles are being added to materials for road resurfacing.

7.12 Warranties

Roof warranties can provide a false sense of security to architects and home owners. The exclusions and limitations set by manufacturers' warranties require careful attention to the fine print in the warranty, to pertinent sections of the manufacturers' specifications manual, and to roofing industry standards. The warranty must be detailed because of the wide variations in roof design and methods of application of materials. The architect, roofing contractor, and owner of the building have the responsibility of design, construction, and maintenance in order to satisfy the conditions of the warranty.

Manufacturers have little or no control over the design of a roof or over the training of the workers who cut and shape the materials, over weather conditions, or over how their materials are used to construct a watertight roof. Instead, the manufacturers fall back on a myriad of specifications to limit their liability. Roof leaks around a ventilator or other roof component that is not properly installed are not covered in the warranty. Although the manufacturer's representative approved the condition of the roof structure as warrantable, the fine print in the document will state that the warranty does not apply when industry standards are not followed exactly. For example, a ventilating fan might come from the factory mounted 15 centimeters (6 inches) from its metal base and flange. Stringent roofing standards and manufacturers' manuals specify that penetrations terminate at a minimum distance of 20 centimeters (8 inches) above the roof surface. Any water leakage would not be covered by the warranty unless the ventilator fan has been installed at the required height.

Perhaps the roof insulation loosens and causes the roofing material to warp and crack. The warranty would not require any action by the manufacturer because its extensive specifications manual and stringent industry standards state that all roof insulation must be firmly attached to the roof deck. The manufacturer could not readily tell how securely the insulation had been attached.

Since the manufacturer of the material has no control over such things as the traffic across the roof by window washers, various service people, or mechanics, water leaks due to this traffic would not be covered by the warranty. Nor is damage covered that is caused by snow, ice, ponding of water, or acts of God, such as high-wind damage from hurricanes or tornadoes. Like fire insurance policies, numerous special endorsements may be needed for common problems, such as a flashing leak endorsement. Written by lawyers, warranties may need interpretation by lawyers.

Manufacturers emphasize that they are not in the business of roof design or installation. Their specialty is materials. There is keen competition between manufacturers' representatives to sell products, and they may go to great lengths to make their products attractive. Warranties against roofing failure are only cost-effective in large complexes such as housing developments, schools, hospitals, resorts, and shopping centers. These have enough at stake to warrant the effort and expense needed to ensure that any dispute over water damages is covered in warranties and can be upheld in expensive and tedious litigation.

What can the average client do to avoid common roof failure? Certainly a quality roofing material from a reputable manufacturer is a good start. A local applicator should be selected from the manufacturer's list of approved roofing contractors. If the installation is extensive enough to make it cost-effective, a qualified inspector could be employed to make sure the architect's plans are satisfactorily carried out and, in case of a warranty dispute, to keep records of the materials installed, details of the methods used, and weather conditions at the time of installation. An inspector will usually cost less than 10 percent of the cost of the roof. Finally, periodic inspection and maintenance should be made by a contractor approved by the manufacturer's representative. Most manufacturers take pride in the performance of their products and will readily cooperate.

7.13 Sealants

Adhesives and sealants are usually discussed together, but there are differences. Adhesives are intended to hold surfaces together;

sealants are intended to exclude or contain substances. In surface preparation, formulation, and application, adhesives and sealants have much in common with one another and with paints.

Sealants must have low viscosity so that they can be extruded or poured, yet they must harden to form a bond with the substrate, not flow under stress, and not crack or leak. The purpose of a sealant is to prevent the passage of air, water, and heat through the joints and seams on the exterior of buildings.

The universal sealant will probably never be found since there are many different properties needed for the varied uses and environments. The problem is to determine which properties are the most important in the selection of the sealant for a particular application. Understanding the essential properties needed for the performance of a sealant and the varied nature of the common building sealants on the market is necessary for the best selection. In practice the selection of the sealant for a particular application becomes a tradeoff between the cost of the sealant and the properties that are most desired.

In view of the continued research and development of modified and improved acrylic sealants, the specific needs of the architect, the specifier, and other professionals are probably best met by their keeping in close contact with the manufacturers. Information would then become available regarding the evaluation of new products designed for the specific requirements of the building industry.

7.13.1 Movement capability

The movement capability of a sealant in a joint is one of the most important properties to evaluate when determining expected performance. This capability is the maximum extension or compression (compared to its original dimension when installed) that a sealant can make without experiencing bond failure. In other words, it is the maximum percentage of change in the width of the joint that the sealant can tolerate. The percentage is obtained by dividing the change in the width of a joint by the original width, expressed in percent. The movement capability is a unique parameter of the sealant that involves many of its physical properties. Movement capability is rated as ±5 percent, ±12.5 percent, ±25 percent, or ±50 percent. The plus indicates the maximum extension; the maximum compression is indicated by the minus sign. (This and other tests are described in ASTM C 719 and ASTM C 920.) Note that a given thermal movement of a panel is a larger percentage of a small joint than it is of a larger joint. Thus the small joints require a larger percentage of movement capability of the sealant. This means that there is a practical limit to how narrow a sealed joint can be and a limit to how wide a sealed joint can be.

There is a continual challenge to create sealants with ever larger movement capabilities. Much effort goes into joint designs to reduce the demand for higher movement ability. Another way to avoid stress in the sealant and loss of bonding to the sides of a joint is to cover the bottom of the joint with a strip of polyethylene or other release agent. As the sealant ages and the solvent escapes, the sealant shrinks, hardens, and the movement capability slowly decreases.

7.13.2 Hardness

The hardness of sealants is measured with a Shore durometer. This standard instrument measures the depth of a substance penetrating a joint. The pencil-type Shore A penetrometer measures the penetration of a blunt probe into the test material on a scale from 0 (no penetration) to 100 (complete penetration). Acceptable values for sealants range from 15 to 50.

7.13.3 Adhesive promoters

High-performance sealants require adhesion to aluminum, steel, and glass in addition to porous materials such as wood and concrete, where the mechanical adhesion is relatively easy to obtain. The search has resulted in phenolic derivatives being incorporated into sealants as adhesion promoters.

The discovery and development of silanes provided an improved adhesion promoter. (Silanes are compounds of silicon and hydrogen that are used to bond organic polymers to inorganic surfaces.) Silane has an adhesive functional group at each end of the molecule. One end attaches to itself, whereas the other end forms an adhesive bond with the hydroxyl sites on glass, metals, and masonry.

7.13.4 Primers

Primers are needed in certain applications. The highly elastic sealants normally do not wet some surfaces and, therefore, require a primer. For example, the polysulfides require a primer for glazing, the urethanes require a primer for glass and metals, and silicones require a primer for concrete. For those cases where even silane additives in sealants do not provide the desired bonding, such as with stainless steel and anodized aluminum, it is necessary to use primers.

There are two types of primers. One type supplies a monomolecular film, whereas the other type automatically produces a film in place. The monomolecular film provides a layer one molecule thick. This is the ideal adhesive following the principle that the thinner the layer, the better its performance. These monomolecular film primers are dilute solutions of a silane in an anhydrous solvent, such as toluene or xylene. Before application, the surface must be wiped clean with a solvent-soaked lint-free cloth to expose the active hydroxide centers on the surface. The silane solution should then be wiped on the surfaces with a clean, lint-free towel and the solvent allowed to evaporate. The sealant must be applied within 15 minutes, before the exposed end of the molecule has a chance to react with vapors in the air. Although this method of making a truly monomolecular film is questionable, the improved adhesion performance warrants the effort.

7.14 Sealant Types

Introduced as the first elastomeric sealant for use in modern curtain-wall construction, polysulfides easily replaced oil-based caulks and were an immediate success. However, with the discovery of even higher-performance sealants, the popularity and use of polysulfides has declined.

In this scientific age no product or building material can expect to retain its ranking as the best for long. The slogan of science and technology is that it is only necessary to conceive of miracle materials in order to create them. To keep up with progress, technology must be presented with a sound, scientific foundation.

7.14.1 Oil-based caulks

Oil-based caulking compounds are prepared from a variety of natural oils, such as linseed oil. They may contain fillers, catalysts, solvents, and plasticizers. These oil-based caulks have a fair adhesion to wood, but should be used with a sealant primer to prevent the oil they exude from staining the surfaces. An inexpensive caulking or glazing material, much used in the past, is made from clay and fiber-filled linseed oil. A catalytic curing agent causes the linseed oil to harden in place due to oxidation of its double bonds with oxygen from the air.

With a movement capability of only a few percent of the joint width, these caulks are mostly sold over the counter to the general public for home repairs. Oil-based caulks are readily available, inexpensive, and easy to use, but their short lifetime and marginal performance makes them of questionable value.

7.14.2 Polysulfide sealants

Polysulfide sealants, based on a form of rubber, are prepared by pouring dichloroethylformal into sodium polysulfide (a solution of sulfur in sodium sulfide) in the presence of an emulsifying agent. They are purchased in two parts and cure when mixed because of the lead-dioxide catalyst in one of the separate containers.

Polysulfides have the distinctive odor of rotten eggs. A small amount of this odor is retained in the sealant after curing. Efforts to mask the odor with perfumes are unsuccessful.

Polysulfide sealants are modified with carbon black, which acts as a reinforcement agent and achieves resistance to weathering by shielding the polymer from radiation. There are plasticizers which increase flexibility and silanes to improve adhesion. Calcium carbonate is added as a filler to provide bulk and reduce the cost. However, such a filler hastens deterioration of the sealant by increasing water penetration. The addition of trichloropropane causes cross-linkage in the polymer, thereby decreasing its flexibility, but increasing the modulus of elasticity and recovery ability of the sealant. The resistance of the sealant to jet fuel and its excellent flexibility at low temperatures make it popular in the aircraft industry.

7.14.3 Butyl sealants

The first butyl sealants were based on a solution of butyl rubber or chlorobutyl rubber in mineral spirits (petroleum distillate). These polymers are made from the copolymerization of isoprene with isobutylene or chlorobutylene. Because of their movement capability, butyl sealants rapidly replaced the polysulfide sealants when butyl rubber became available in the 1950s. Low-cost butyl sealants have good stability and are resistant to water and organic solvents.

Calcium carbonate and talc are used as fillers. Carbon black can be used as an additive that serves both as a reinforcement agent and as a stabilizer to shield the polymer from the effect of the sun. Butyl sealants contain adhesion promoters such as silane, talc, calcium carbonate fillers, and a thixotropic agent to reduce slumping in the joint when freshly applied.

One of the properties of butyl sealants that makes them superior to the polysulfides and competitive with the acrylic sealants is their +12.5 percent movement capability. These butyl sealants have good water resistance and good adhesion formulation, so they do not require a primer. Their stickiness, until they skin over, causes them to pick up dust from the air. They become stiff in cold weather and soft in hot weather. The polymers in butyl sealants are already cured. They do not harden or deteriorate due to continued polymerization by ultraviolet radiation in sunlight. These one-package solvent (naphtha) sealants are plasticized with small amounts of low-molecular-weight polybutene oils.

If naphtha (mineral spirits) is used as a solvent, its evaporation over time produces a potential shrinkage in volume of about 40 percent. This can be expected with an accompanying amount of hardening. If enough of this polybutene oil plasticizer is added, a permanently soft mastic is formed, which is suitable for sealing curtain-wall joints. Over the last few decades extensive research and development has gone into the application of butyl rubbers as sealants in the building industry. In spite of the superior performance of the silicones and urethanes as sealants, the butyls have a secure place as glazing tapes, used to fit glass into window frames, and as hot-melt sealants.

A typical formulation of a solvent-based butyl sealant might contain (by weight) 15 percent butyl resin solids and as much as 50 percent of a filler of talc and calcium carbonate, with the remainder consisting of a plasticizer and other additives. In such a formulation the shrinkage from the evaporation of the solvent could be as high as 40 percent in volume.

Home remodelers, guided only by the claims that are printed on the gun-ready cartridge and the easy cleanup of water-emulsion products, frequently experience improper selection and use of the product. The choice of a sealant is made especially confusing when "with silicone" is included in the product's name without any indication of the amount of silicon the product contains.

7.14.4 Acrylic sealants

The latex-emulsion acrylic sealants got their start in the 1960s when latex-emulsion paints became popular. Due to their lower cost, easier cleanup, and good performance, acrylic sealants have largely replaced the solvent-type acrylics and are in close competition with the butyl sealants. The acrylics are water emulsions of a formulated acrylic in a small amount of water with detergent as an emulsifying agent.

Acrylics have good bonding ability and adhere to a wide variety of surfaces, so they do not require a primer. Their long shelf life is an advantage. These one-component sealants are fast skinning, tack-free in about 30 minutes, and can be painted over in one or two weeks. Because these sealants are emulsions in water, they should not be applied when rain or freezing weather is expected. They are fairly resistant to ultraviolet light from the sun, especially when protected with additives, and perform well as exterior and interior sealants. They do not require adhesive additives because of their inherent good adhesion.

The maximum movement capability of acrylic sealants is +12.5 percent of the joint width. However, loss of water over a period of time causes these sealants to harden and lose some of their moderate movement capability. Because of their water content, there is some shrinkage and an increase in hardness that occurs over time. They are somewhat flexible, but have poor elastic recovery.

Like the polysulfide sealants, the acrylic sealants have a characteristic odor which makes them less desirable for indoor use. They are used for light construction and are readily available in retail stores to the cost-conscious do-it-yourself home owners. These popular sealants constitute a thriving market.

7.14.5 Urethane sealants

Urethane sealants rank second among the sealants used in industry. A chief component is a polyurethane. The polyurethanes are formed by the reaction of an isocyanate group with an alcohol or an amine. In the reaction of this polymer with water, CO_2 is formed. Various organic isocyanates can react with the various alcohols or amines to provide a large family of polyurethanes. These reactions of the isocyanate group make the number of potential sealants almost unlimited.

Two-component urethane sealants require the thorough mixing of two packages. Package A contains primarily a low-molecular-weight polyurethane with some remaining unreacted isocyanate groups, some fillers, and plasticizers. The A package may also contain small amounts of sealant modifiers such as colorants, thixotropic agents (to avoid flow after application), solvent, ultraviolet absorbers, and adhesion additives. Depending on the application, silane adhesion promoters may either be added or used as primers. Package B contains the alcohols, amines, or other isocyanate reactants. There is a problem with the slower rate of curing in the two-package system. It can take several days for the surface to lose tack and skin over in cold dry weather.

The one-package ready-to-use sealant requires special care to prevent the reaction of moisture with isocyanate groups to form carbon dioxide. Its properties compare favorably with the two-package urethane systems, but the short shelf life is a problem.

7.14.6 Silicone sealants

Silicone sealants rank first among the sealants used in industry and are most used in high-rise buildings. Their high performance, resistance to low temperatures, high ozone resistance, good adhesion to surfaces, and very high movement capability continue to make them the favorite product, regardless of high cost. Silicone sealants cure on contact with the moisture in the air to form acetic acid, which provides their characteristic odor. Their availability in a one-cartridge container and their reasonably good shelf life make these sealants popular with industry and the general public.

The usual formulations of silicone sealants provide long service in exposure to harsh environments, good stability, high peel and tear resistance, and low shrinkage and weight loss. Silicones have a natural resistance to weathering due to the stability of the silicone polymer. They have set another record with the maximum movement capability increased to $+100$ percent extension and -50 percent compression. The constant need for greater movement capabilities, with little or no increase in hardness with time or temperature, promises to retain this sealant's top-ranking performance and popularity for many years to come.

Silicone sealants are made in three grades of modulus. High-modulus silicone sealants contain large amounts of fused silica as a reinforcement agent and provide maximum movement capability. Medium-modulus silicone sealants are prepared from a lower-molecular-weight silicone polymer with the addition of a small amount of toluene, fused silica, and calcium carbonate. Low-modulus silicone sealants are a blend of silicone polymers, including a straight-chain polymer, calcium carbonate, fused silica, and a plasticizer to provide a softer, highly extensible sealant.

A common problem is the formation of a low-molecular-weight cyclic silicone oil, which can migrate to the surface of the sealant and stain light-colored building materials such as marble and concrete. The oil may form an undesirable film on glass and metal surfaces. One solution is to protect the surface material with a primer, which forms a barrier coating.

The development of a water-emulsion silicone sealant makes application, paintability, and cleanup much easier. With no sacrifice in performance or longevity, this is an outstanding achievement. This development, again, shows the versatility of silicones and their increasing popularity with people in the building industry and the do-it-yourself retail market.

7.15 Fire Protection

It was the disastrous fire of 1666 in London, England, that resulted in new building codes and municipal ordinances banning thatched roofs and wooden chimneys. But there was still much to learn. It has been claimed that San Francisco was damaged more by the fire that followed the 1906 earthquake than by the earthquake itself. During World War II it was the fires as much as the bombings that destroyed Dresden, Germany. In Los Angeles, Calif., one of the authors (LRB) was instrumental in the banning of backyard incinerators as fire hazards and smog producers.

Much attention is devoted to the causes and prevention of fires, a major cause of loss of life in the civilized world. This attention is directed toward efforts to understand the principles and methods of fire prevention. By giving careful consideration to the choice of building materials and their best use, we can do much to confine a fire and prevent its spread.

The thermal and mechanical properties of common building materials are basic to fire prevention and confinement. With the aid of computer models, the relative fire resistance and fire rating of different building materials can be evaluated.

7.15.1 Materials at high temperatures

The physical properties of building materials determining their resistance to a fire are thermal conductivity, specific heat, and diffusivity. Thermal conductivity, also known as heat conductivity, is defined as the quantity of heat that passes in unit time through a unit area of material of unit thickness when its opposite surfaces differ in temperature by 1 degree. Specific heat is defined as the amount of heat required to raise the temperature of a unit mass by 1 degree. Thermal diffusivity is obtained by dividing the thermal conductivity by the specific heat per unit volume.

The ASTM standard temperature-time relation specifies how rapidly the temperature should rise in a model fire. For example, the standard requires that, starting at room temperature, the temperature in a model fire should reach 1000°C (1832°F) in about 1.5 hours.

Usually a fire starts with the combustion of a single object in a room. This is followed by a growth period and, finally, by a "flashover" as the fire goes into the full development phase. During the growth period, the evacuation of occupants should be completed.

The risks to people and of early structural failure begin with the rapid increase in temperature and the resulting sustained burning. The "severity" of a fire is measured by the effect on the boundaries of the burning room, rather than the temperature-time exposure of the room.

The nature of the fire depends on the fuse—whether it is wood, wallboard, or concrete. Thus to predict the ability of a room's boundaries to confine the blaze, the nature of the individual materials making up the boundaries must be considered. This requires a knowledge of the effect of the resulting high temperatures on these common structural materials. Temperatures can be expected to range from 800°C (1472°F) to a maximum of about 1200°C (2192°F).

In severe fires all common building materials, such as wood, concrete, and steel, suffer structural damage to a variable extent. Depending on the nature of the materials, they may char or burn, fragment or disintegrate, or expand, warp, or melt.

Thermal "diffusivity" measures how readily heat can spread through a material. Steel has a much higher value than concrete or wood, causing steel to be particularly vulnerable to the heat of a fire. The fire resistance of materials can be measured by the length of time the materials remain functional when exposed to standard fire-test procedures as specified by ASTM E 119.

The time needed for any material to reach a given temperature can be extended by providing thermal insulation for the structural materials.

7.15.2 Wood exposed to fire

As the temperature of wood gradually rises during a fire, combustible vapors are emitted by the decomposition of the surface. These vapors ignite spontaneously when the temperature reaches about 400°C (752°F). The ignition temperature of the wood depends on several factors, including the species of wood, porosity, density, and any surface coating. After ignition, the surface layers become charred. The wood char, with a lower thermal conductivity because of its reduced moisture content, acts as a protective coating. The rate of charring continues at a slower rate until the remaining core of the wood can no longer support the load it is bearing.

Although wood can ignite and burn if it is thin or splintered, thick timber beams tend to char slowly at a uniform rate. The average rate of charring for common types of wood is usually assumed to be about 4 centimeters (1.4 inches) per hour when subjected to the "standard fire exposure." The rate of charring increases as the temperature rises and is much higher if the end grain of the wood is exposed. The surface char, once formed, provides thermal insulation for the underlying wood. At the same time, the loss of moisture causes the specific heat of the wood to increase slightly.

7.15.3 Concrete exposed to fire

The relatively low thermal conductivity of concrete is decreased by about 25 percent when heated to 800°C (1472°F) due to loss of moisture.

Creep is not significant until the temperature reaches 400°C (752°F), but creep usually has little effect on the performance of concrete.

Concrete loses about half of its modulus of elasticity at 400°C (752°F) and does not recover as it cools.

Between 300 and 600°C (572 and 1112°F) the small amount of impurities of iron compounds in the concrete change color. The concrete changes from the usual gray to pink; then it becomes red as the temperature rises. The color of the concrete bleaches to white at 800°C (1472°F). During a fire, this gradual conversion of iron impurities in the aggregate to ferric oxide from 300 to 600°C (572 to 1112°F) and the conversion of iron oxides to a colorless compound at around 800°C (1472°F) serve as convenient indicators of the maximum temperature reached by the concrete.

The loss in compressive strength of concrete depends on both the temperature reached and the duration of each heating period. As might be expected, the residual strength is further reduced on cooling, often losing as much as 50 percent of the strength the concrete had when still hot. This loss in strength can seriously affect the reusability of concrete structures.

7.15.4 Steel exposed to fire

Cold-rolled steel, used in structural steel, decreases rapidly in strength around 400°C (752°F) as the temperature of the fire increases. Cold-drawn wire (used in prestressed concrete) loses strength even faster. The yield strength used in the design of steel structures at working loads (the failure point on a stress-strain curve) decreases to about half its ambient value at fire temperatures above 600°C (1112°F).

Creep, a deformation which takes place gradually over a period of time, also occurs at an increasing rate as the temperature rises. Creep starts to be a problem above 500°C (932°F), depending on the load carried by the support. The resulting deformation can eventually cause failure of the structure.

The high thermal conductivity and thermal diffusivity of steel, about 100 times larger than that of wood or concrete, can be a serious problem in a fire. The warping and distortion of steel supports, caused by their thermal expansion and contraction as they cool, can weaken a structure seriously or lead to its collapse. Consequently, load-bearing structural steel requires special precautions to prevent failure of a structure resulting from the high temperatures that might be reached.

7.15.5 Polysteel forms

Along with dozens of alternative building materials designed to lower the amount of wood used in building a home, polysteel forms seem to have advantages worth considering. They are fire-resistant, energy-efficient, nearly soundproof, moistureproof, termite-resistant, and affordable, and they may command lower insurance rates. The method of construction involves placing hollow polystyrene forms over steel rods, stacking them on top of one another, then slowly pouring cement inside. When the cement dries, there is a solid wall of 15 centimeters (6 inches) of concrete surrounded by the insulating polystyrene with no cracks, no hollow cavities, and no spaces. Variations of this method have been around since 1930.

7.16 Fire Security

Building codes and fire insurance ratings are based on the performance of materials, not their composition and structure. For example, local building codes may specify that the "fire-resistive" load-bearing members, such as walls, beams, and columns, have a fire-resistance rating of 3 or 4 hours.

Fire resistance is the ability of a material to withstand exposure to fire for a specified period of time without the loss of load-bearing capacity or the ability to serve as a fire barrier. The calculation of the time is usually complex, even after the necessary laboratory fire-resistance tests have been carried out. Such an involved process is best done by engineers trained in this field.

Fire protection is one of the prime design features of buildings. The Universal Building Code is frequently incorporated directly into local municipal building codes. In case new building materials, such as foamed and reinforced plastics, are not covered in building codes, they probably would be used under supervision of the local municipal building administrators.

7.16.1 Fire detection

Fire alarms and smoke alarms are essential devices. In multistory buildings extending above the reach of municipal fire-fighting equipment, all rooms and storage areas should have automatic water sprinklers. Modern fire extinguishers should be available in high-risk areas on each floor.

In commercial buildings heat sources, such as electric coffee makers, sandwich grills, and kitchen appliances, could be fire hazards if they are used near flammable materials. Smoking should be banned in the building or confined to fire-protected areas.

7.16.2 Containment of fire

The difficulty of confining a fire to the area where it started depends on the resistance of the materials used to construct the ceiling, walls, and floor. These interior elements should confine the fire until the occupants can be safely evacuated. Fire doors, if in frequently used passageways, should have automatic closures activated by smoke or by heat detectors to help confine the fire. The automatic closure of ventilating ducts can retard the spread of smoke and fire between floors. Earthquake and other disaster-prone areas require fire-door closures that are activated by vibrations or shock waves.

Gypsum board, used in wall panels and ceilings, is not only a good thermal insulator, but the calcium sulfate composite absorbs heat as it is dehydrated at the high temperatures reached by the fire.

Inorganic compounds, such as calcium carbonate or calcium oxalate, might be considered for use in an organic protective surface coating. Heat would be dissipated both in the thermal decomposition of the compound and in the intumescent fragmented particles detached from the surface by the sudden gas formation.

By coating the interior surfaces of a room with these intumescent materials, the high surface temperature can be reduced as these chemical materials swell and decompose. When heated, intumescent materials form foams or nonflammable gases, or they disperse fragmented particles. As these gases or particles spontaneously leave the wall or ceiling surfaces, they take heat with them, thus cooling the hot surface.

Fire-retardant glass panels can be used in skylights, windows, doors, interior walls, and store fronts. The transparent glass units, made from two or more lites (sheets) of tempered glass, are separated by a gel of polymer salt and water, which crystallizes to a solid when heated. This fire-rated glass assembly can delay the spread of fire for 20 to 90 minutes.

7.16.3 Protection of structural elements

Mammoth modern structures depend on steel beams and girders for structural support. Yet steel, with its high thermal conductivity, readily absorbs the heat produced in a fire, thus causing its temperature to climb rapidly. Reinforcement bars embedded in concrete floors and the steel wires and cables used to prestress concrete slabs are vulnerable to excessive heat.

Common design measures to protect the steel supports involve the concepts of insulation, isolation, and individual cooling of the structural steel supports. Insulation can involve both protection from heat and radiation protection from burning materials. Structural steel is usually protected with some kind of insulating and reflective coating. The U.S. Steel Corporation has pioneered isolation and cooling of the supporting columns of a building by successfully demonstrating the use of hollow columns of weathering steel purposely placed on the exterior of buildings. Cooling water, containing an antifreeze and a rust inhibitor, circulates through the columns.

A structural steel framework should have special fire protection. Heat shields and fire walls can be used. The steel columns can be isolated in utility shafts provided with good ventilation. Or the building can be designed to use an exterior steel support system out of thermal contact with the walls of the building. Gypsum board panels provide both thermal insulation and absorption of thermal energy by the dehydration of the gypsum. These are only a few of the cost-effective methods of protection that are available.

Each structure poses a different challenge with its unique style, the environment, and its intended use. The goal is to provide the maximum time for the evacuation of personnel and for fire fighters to confine, control, and extinguish the fire. This makes each obstacle in the propagation of the fire worthwhile, even if it only provides a temporary delay in the spread of fire.

7.17 Repair of Damage

Over the years it has seemed natural to tear down "obsolete" or damaged buildings to make way for "progress." The old buildings were replaced with "modern," taller buildings with up-to-date features and more efficient allocation of space. When fire damage was severe enough that the insurance could cover much of the expense, demolition and rebuilding were to be expected, especially with the chance to modernize and increase the floor space of the building. But attitudes are slowly changing, and the feasibility of making repairs should be considered.

Repairing fire-damaged buildings becomes more attractive now that there are concerns for the environment, shortages of municipal and private sites that accept used building rubble, and our diminishing natural energy and building materials resources.

Information concerning temperatures reached in the various areas damaged by a fire can be obtained by collecting samples of glass, wood, plastics, and metal pieces with distorted shapes. Since metal alloys, plastics, and glass soften over a range of temperatures before melting, a careful inspection of such materials can give considerable information. Handbooks can be consulted to find the softening points and distortion melting temperatures of common building materials.

The temperatures reached and their duration in the various areas affected by the fire are needed before an assessment of the severity of the damage can be made. A detailed examination of damages may involve removal of smoke stains that might conceal discoloration and cracks in the walls and ceiling where the fire was the hottest.

If the fire did not get out of control and the damage was confined to a relatively small portion of the building, the next step is to determine the condition of the structural supports. The feasibility of repairing buildings depends on the assessment of the severity of the fire.

Buildings with a wooden framework must be inspected to find evidence of charring or splintering of structural wood. This involves determining the location and depth of charring to learn whether the structure is still up to specifications and whether it contains a reasonable margin of safety before the building can be satisfactorily repaired.

Concrete structural elements, if seriously damaged, may show distortion, sag, or deep spalling. Or the concrete may be buckled or deep transverse cracks may be found that could be suspected of seriously weakening the structure.

Steel structures may show signs of heat breaching their heat shields. If the steel beams are twisted or out of line, serious damage is apparent.

There are several other ways to determine the extent of the damages. The records of the fire department will note the duration of the fire and any difficulties encountered in extinguishing the fire in various areas to which it had spread. Another source of information is the company that provided the fire insurance coverage. The reports of competent investigators employed to determine the extent of the damages should be informative. And the insurance company should be willing to share the results of their inspections with the professionals who are considering the repairs.

The extent of all this damage will determine the feasibility of repairs, keeping in mind that the structure will have to pass the same rigorous standards of safety as a new structure. It is possible that test loading of the structure will be needed to determine whether the structure has a sufficient margin of safety to warrant repairs. Hopefully the fire-detection devices were effective and the fire was confined and extinguished before any irreparable damage occurred.

7.18 Health Hazards

According to the insurance companies, roofers are steeplejacks. Injuries to roofers are among the top three in frequency and are the most harmful. Falls from roofs or scaffolds are not the only hazards. Hot asphalt emulsions can cause serious skin burns; they stick and continue to cause damage. Long-sleeved shirts should be worn to protect the skin from splashes of hot asphalt and from the cancer-causing ultraviolet rays of the sun. OSHA rules are very strict and specific. Failure to comply can result in injury, destruction of the building, and expensive fines. These requirements are updated frequently, and the amounts of fines are increased.

The solvents used in sealants may include chlorinated hydrocarbons (toxic to the liver) and aromatic solvents (such as benzene and toluene) that are carcinogenic. The amines used as curing agents for the epoxides may cause dermatitis of the hands and face and, if inhaled, serious respiratory problems.

Most polymers are made from toxic monomers. When polymers are heated in a fire, especially linear polymers, they decompose to release their monomers. Polyurethane foam burns readily to produce hydrogen cyanide (a toxic gas) along with a dense, black smoke. Some polymers such as phenol formaldehyde, which is used in wallboard and carpets, are suspected of releasing formaldehyde by slowly decomposing at room temperature. Over an extended period, formaldehyde may be a serious health hazard. In considering toxic substances, the concentration and the length of exposure are both of vital importance.

If polymeric materials are disposed of by incineration, the noxious gases are respiratory irritants and are toxic if inhaled over extended periods.

Because burning materials often emit caustic and deadly gases, more deaths from fires are caused by smoke inhalation than by burns.

Instead of waiting until new chemicals are shown to be toxic, we should take precautions to avoid unnecessary exposure to chemicals in the workplace and at home.

Terms Introduced

Aerosol	Emulsions of water in oil
Asphalt derivatives	Foamed plastic membrane
Asphalt emulsion	Mineral-fiber shingles
Asphaltic bitumens	Polysteel forms
Blown asphalt	Sealant adhesive properties
Colloidal systems	Sealant movement
Cracked asphalt	Sealant primers
Cut-back asphalt	Single-ply membranes
Emulsions of oil in water	Wood shingles

Questions and Problems

7.1. What properties are most desirable when selecting roofing materials?
7.2. What is blown asphalt?
7.3. What is an asphalt emulsion?
7.4. List common problems in roof design and construction and their solutions.
7.5. What is the purpose of a flashing?
7.6. Discuss types of insulation and their importance.
7.7. What is the difference between adhesives and sealants?
7.8. Discuss a common sealant and its properties.
7.9. Discuss movement capability and its importance.
7.10. Why are sealant primers needed?
7.11. Why is the expansion coefficient of the substrate important in the selection of a sealant?
7.12. What precautions can be taken to restrict a fire to the starting area?
7.13. How is the temperature of a fire estimated from the appearance of the remnants?
7.14. How can one determine the extent of fire damage?
7.15. How can structural steel be protected from damage by fire?
7.16. Why is smoke inhalation so dangerous?

Suggestions for Further Reading

Ronald Brand, *Architectural Details for Insulated Buildings,* Van Nostrand Reinhold, New York, 1990.

T. T. Lie, ed., *Structural Fire Protection,* American Society of Civil Engineers, New York, 1992.
B. Harrison McCampbell, *Problems in Roofing Design,* Butterworth Architecture, Boston, 1991.
William C. McElroy, *Roof Builder's Handbook,* Prentice-Hall, Englewood Cliffs, N.J., 1993.
Robert X. Peyton and Toni C. Rubio, *Construction Safety Practices and Principles,* Van Nostrand Reinhold, New York, 1991.
Don A. Watson, *Construction Materials and Processes,* McGraw-Hill, New York, 1986.

Chapter 8
Glass

8.1 Introduction — 254
8.2 Glass Formation — 254
8.3 Properties of Glass — 255
   8.3.1 Viscosity — 255
   8.3.2 Glass-Transition Temperature — 256
   8.3.3 Nucleation and Crystallization — 256
   8.3.4 Tensile Strength — 257
8.4 Basic Silica Structures — 257
   8.4.1 Amorphous Solids — 258
   8.4.2 Crystalline Silica — 258
   8.4.3 Silica-Type Structures — 258
8.5 Kinds of Glass — 259
   8.5.1 Soda-Lime Silica (Soft) Glass — 259
   8.5.2 Plate Glass — 260
   8.5.3 Borosilicate Glass — 260
   8.5.4 Lead Glass — 260
   8.5.5 Optical Glass — 260
   8.5.6 Photosensitive Glass — 261
8.6 Safety Glass — 261
   8.6.1 Wired Glass — 261
   8.6.2 Tempered Glass — 261
   8.6.3 Chemically Toughened Glass — 264
   8.6.4 Laminated Glass — 264
   8.6.5 Special-Purpose Laminates — 265
8.7 Insulated Glass — 265
8.8 Solar-Control Glass — 267
   8.8.1 Tinted Glass — 267
   8.8.2 Coated Glass — 269
8.9 Structural Glazing — 269
8.10 Glass Fibers — 270
8.11 Glass Ceramics — 271
8.12 Recycling — 273
Terms Introduced — 273
Questions and Problems — 273
Suggestions for Further Reading — 274

8.1 Introduction

One of the oldest building materials, glass dates back 5000 years to our earliest recorded history. Probably the first glass was an accident or was produced by volcanic action. Later, with creative skills, glass became valued as ornaments and containers.

The first known producers of glass were the Egyptians; then production moved to Venice. The invention of the glass blower's pipe in the first century B.C. allowed glass to be heated to a higher temperature and then blown and shaped. After the third century A.D. glass was used in daily life. During the fifteenth century changes began when the English invented the lead crystal, highly prized for its brilliance. The exact composition of glass was kept a secret until the beginning of the twentieth century, when machine production began. This resulted in advances in science and technology and increased research.

The preparation of glass, its unique physical and chemical properties, the structure that accounts for the unusual nature and versatility that make glass indispensable, its uses, its care, and recycling are described in this chapter.

8.2 Glass Formation

The conventional method of making glass is to cool a molten mixture of silicates so rapidly that it does not have time to crystallize. This method of formation is the reason why glass is known as an undercooled liquid.

Because of varied compositions and the complexity of structures, it was only recently that advances in instrumentation have made possible fundamental research in glass. By using modern techniques, glass can be formed by a wide variety of methods, such as vapor condensation, precipitation from solution, cooling molten mixtures under high pressure, and high-energy radiation of crystals.

Consequently, the definition of a glass has been modified by some scientists to describe its characteristic properties rather than its method of formation. In addition to the usual properties of solids, such as rigidity, hardness, and brittleness, these properties include transparency, high viscosity, and lack of an ordered large-scale crystalline structure.

Since the silicate glasses are by far the most widely used glasses, they will be emphasized in this chapter. Thus, the ASTM definition of glass: "An inorganic mixture that has been melted and cooled to a rigid condition without crystallizing."

A common use of glass is in flat sheets for windows in buildings and automobiles. Sheet glass is prepared by molten glass passing between water-cooled rollers as it cools. The surface roughness can be removed by grinding and polishing to make plate glass. However, this process has largely been replaced by the float glass process, in which the molten glass flows from the furnace so that it floats along the surface of a bath of molten tin. The glass assumes the perfect surface of the molten tin as it passes over. At this temperature the molten tin forms a powdery coating of tin oxide by reacting with oxygen in the air. To prevent this oxide formation, the air over the molten tin is replaced by inert nitrogen gas. The temperature is reduced gradually as it moves along until the sheet glass is hard enough to be fed into rollers without marking the under surface of the glass. During this process, internal stresses are set up in the glass. These are relieved by the annealing process, in which the glass is reheated to its annealing temperature for about 15 minutes or until the internal stresses in the glass are substantially relieved. Then it is cooled gradually to room temperature at a predetermined rate to produce an annealed glass that is stronger, more uniform, and without distortions, strains, or other imperfections.

The selective transparency of glass to the visible portion of the sun's spectrum accounts for its extensive use in windows. With energy at a premium, the many ways that glass is modified to transmit or confine the desirable portions of solar radiation have produced many innovative products. Its fragility and crack behavior creates problems in homes and in industrial use unless it is tempered, especially where it is in frequent bodily contact with people.

8.3 Properties of Glass

Silicate glass has the properties of an undercooled liquid. It can sag out of shape if not supported in a vertical position. For example, a sheet of window glass in a horizontal position will have an instantaneous elastic extension followed by a creep, the rate of which gradually decreases to a constant value. Another unique property is that glass is slightly flexible and completely recovers from a slight flexure. This property causes a water goblet to ring when thumped.

8.3.1 Viscosity

As glass is heated to the melting point, it gradually becomes thinner and less viscous. Viscosity is a measure of the fluidity of a liquid or viscous substance and is measured in cgs units in poises. (The cgs unit of viscosity is the force in dynes, 10^{-5} neurons, required to rotate a paddle 1 square centimeter in area, 1 centimeter from another identical stationary surface, at 1 revolution per second.) On this scale, water has a viscosity of about 1 centipoise at room temperature. The viscosity of glass at its working temperature is about 10,000 poises. At its softening point, its viscosity is about 40 million poises. Glass has an abrupt change in viscosity at its glass-transition temperature, which is followed by an extremely large change at its melting point. Glass is an ideal liquid since its viscosity does not depend on the rate of shear or the length of time it has been under shear.

8.3.2 Glass-transition temperature

In addition to viscosity, many physical properties of glass, or other substances in the glassy state, show an abrupt change at the glass-transition temperature. These properties include specific volume (the volume of 1 gram), heat, and electrical conductivity. Note that each of these can be measured so quickly that the components do not have time to return to an equilibrium structure. Also, above the glass-transition temperature, when the substance is no longer in the glassy (rigid) state, the viscosity is small enough for molecular movement to begin to occur.

As glass is heated from its glassy state to the glass-transition temperature, its thermal energy becomes large enough to break many of the bonds holding the chains together, suddenly making glass more fluid. Further heating gradually increases its fluidity until the remaining chain-linkage bonds break at the melting point. These changes depend on how rapidly the glass is heated and on its previous heat treatment. For these reasons, glass-transition temperatures are usually approximations.

8.3.3 Nucleation and crystallization

The formation of silicate glass depends on cooling the molten mixture so rapidly that crystallization does not have time to occur. Crystallization takes place in two steps: (1) the formation of crystal nuclei (nucleation) and (2) the growth of crystals on these nuclei. As the molten mixture cools, the rate of nucleation increases to a maximum and, then falls off. As the melt cools further, the rate of crystallization reaches its maximum due to its increased tendency to solidify and the increased number of nuclei present. The presence of nuclei in the form of impurities can dwarf the rate of nuclei formation and is to be avoided. As the viscosity of a melt increases on cooling to room temperature, both nucleation and crystallization become negligible. Thus the cooled glass reaches a metastable condition with the crystallization process frozen for practical purposes.

8.3.4 Tensile strength

The tensile strength of glass is reduced to a fraction of its value if minute cracks and scratches occur. Like adhesion, the strength of a glass system depends on the uniform distribution of stress. Irregularities or surface defects should be kept to a minimum. Surface defects can be reduced by etching the surface chemically or by remelting to reform the surface.

Another treatment that improves strength is to introduce a compressive stress into the surface, as is done with prestressed concrete. Surface strain is produced by rapid cooling, which causes the surface to cool faster than the interior of the glass. Other methods are (1) an ion exchange to incorporate more bulky ions into the surface, (2) partially using a nucleating agent to crystallize the surface by spreading fine crystal nuclei over the molten surface, and (3) etching with hydrofluoric acid, which works by dissolving the sharp edges of fine cracks and the imperfections which act as points of stress concentration.

In the ion-exchange method some of the sodium ions in the surface of a sodium aluminum silicate glass are replaced with potassium ions by submerging the glass in molten potassium nitrate for several hours. A tenfold increase in strength results from the introduction of the larger potassium ions into the surface. If the cooling temperature is too high, potassium ions diffuse into the bulk of the glass, resulting in lower stress concentration on the surface. These treatments are prohibitively expensive in most applications.

8.4 Basic Silica Structures

Although glass has thermoplastic properties like organic polymers, it has a different molecular structure. Rather than a carbon-chain structure, silicate glasses have alternating silicon and oxygen atoms in ringlike arrangements, with an additional oxygen atom linking the rings together into a spatial arrangement. These rings are attached in small submicroscopic clumps to form an amorphous (noncrystalline) randomly ordered structure. The commercial silicate glasses are best treated as modifications of silica since the silica structure is basic.

8.4.1 Amorphous solids

A solid need not have a continuous, completely ordered crystalline structure. For example, the components of a metallic alloy freeze in clumps of crystals, which become cemented together by a eutectic—corrosion often starts at these crystal (grain) boundaries. Organic polymers with their branched chains can only form crystalline regions. Each of these freezes over a range of temperatures.

Thus it is no surprise that the mineral silica has an imperfect crystalline structure and different physical properties. The silicate multicomponent glasses melt over a temperature range starting at the glass-transition temperature and finishing at what is loosely called the melting point. In these glasses the crystal aggregates are too small to be seen, unlike the tiny crystals in metal alloys and cement.

8.4.2 Crystalline silica

To better understand the properties of silica and the silicate glasses, we can create a mental model of crystalline silica. To do so, you might visualize a ring structure of six silicon atoms at each corner of a hexagon, similar to the carbon atoms in a benzene ring. The silicon atoms in the hexagon are held together, with oxygen atoms making an alternating silicon-oxygen ring. Silica also forms a tetrahedral structure consisting of a silicon atom surrounded by four larger oxygen atoms. One of the oxygen atoms in the tetrahedron is used in the ring structure. The remaining three oxygen atoms of each tetrahedron are used to individually link together the silicon atoms of adjacent rings to form the spatial structure of crystalline silica. This hard, brittle crystalline material has such strong silicon-oxygen covalent bonds that it must be heated to 1710°C (3110°F) to break up this quartz structure so that it can soften and melt.

8.4.3 Silica-type structures

The structure of vitreous silica is similar to that of the mineral quartz, except that its structure is so disordered that it is classified as an amorphous (noncrystalline) solid.

Crystalline silicon dioxide, found in nature as quartz, is a white, opaque hard mineral. Most sand consists of grains of silica left from the erosion of granite and other quartz-containing rocks. Silica is also found in nature as large, clear crystals.

Molten silica is a viscous liquid. The melting process consists of breaking many of the silicon-oxygen bonds to form a multitude of submicroscopic crystals in a frozen, disordered array. When the melt is cooled rapidly, there is no time for these tiny crystalline fragments to form an ordered, crystalline structure. In this noncrystalline condition it is called vitreous silica and has a transparent, glasslike nature. Its use is limited to instruments requiring ultraviolet transmissions and for the preparation of computer chips.

The addition of a glass-network modifier to silica, such as the oxides of sodium, potassium, barium, calcium, magnesium, or lead, weakens its tetrahedral structure. It does this by providing nonbonding oxygen and islands of sodium silicate. These break up the silica structure to make a glass with a lower softening temperature.

8.5 Kinds of Glass

The most common commercial use of glass is the manufacture of glass containers. Next in importance is glass used in windows for buildings and automobiles. The least expensive and most common glass is soda-lime silicate (soft) glass.

Sodium and calcium oxides are network modifiers which disrupt the silica structure, resulting in a lower working temperature. Additives, such as oxides of antimony or arsenic, are used to aid in the removal of bubbles during the manufacturing process. Alumina may be added to reduce weathering and the tendency to crystallize. By adding borax, a borosilicate-type glass is produced with a lower thermal expansion coefficient which, when cooled rapidly, reduces its tendency to crack.

The two main types of industrial glass in common use are soda-lime (soft) glass and borosilicate glass. These are clear, hard, brittle amorphous solids. Soda-lime glass, the most common, consists of a basic mixture of sand, soda ash, and lime. The addition of small amounts of magnesium oxide reduces its tendency to crystallize, whereas a small amount of alumina increases its durability.

8.5.1 Soda-lime silica (soft) glass

The soda-lime silica glass, often referred to as soft glass, has a softening point of about 600°C (1112°F). Soft glass has the approximate composition of 17 percent sodium oxide, 10 percent calcium oxide, and 73 percent silicon dioxide. The impressive drop in its softening point, compared to that of silica, is due to the high percentage of the two network modifiers, sodium oxide and calcium oxide. In addition to the lower softening temperature, soft glass is slightly more flexible, but has lost much of its resistance to thermal shock. This can be partially overcome by prestressing the glass during cooling.

8.5.2 Plate glass

Plate glass has the same composition as window glass (soda-lime silica) and differs from it only in the method of manufacture. Primarily the differences are (1) the longer time of annealing (3 or 4 days), which eliminates the distortion and strain effects of rapid cooling, and (2) the intensive grinding and polishing, which remove local imperfections and produce a bright, highly reflective finish.

8.5.3 Borosilicate glass

To reduce the thermal coefficient of expansion that is responsible for breakage due to sudden heating or cooling of soft glass, some of the sodium is replaced by boron (another network former) in the form of boric acid or borax. This property of reduced thermal expansion makes oven glassware and cooking utensils possible—originally under the trade name of Pyrex.

Borosilicate glass has the approximate composition of 13 percent boric oxide, 80 percent silicon dioxide, 4.5 percent sodium oxide, 2 percent aluminum oxide, and 0.5 percent potassium oxide. Its slightly higher softening point of about 800°C (1472°F) is offset by improved resistance to thermal shock. Due to improved chemical resistance, as well as reduced thermal expansion, borosilicate glass has found many new household and industrial uses.

8.5.4 Lead glass

The lead and potassium contained in lead-potash silica glass are in the form of metallic oxides which combine with molten silicon dioxide to form high-molecular-weight silicates. As more lead is used, it not only adds to the weight of the glass, but increases its refractive index, causing cut glass to have a characteristic sparkle.

8.5.5 Optical glass

The refractive index of glass is adjusted in optical glass to give a wide range of values. It is needed for bifocal and trifocal eye glasses, telescopes, and other optical instruments. Annealing the glass relieves mechanical strains formed as the surface cools before the interior. For this reason, and to reduce optical distortion, the glass must be cooled gradually.

8.5.6 Photosensitive glass

By incorporating tiny crystals of chlorides of copper, silver, or gold into a molten glass, brief exposure to sunlight produces a temporarily darkened glass as the chloride is decomposed to form the metal and chlorine. However, unlike the latent image formed by light on silver halides suspended in gelatin in a photographic film, the chlorine atom has nothing to combine with chemically inside the glass. Therefore, the metal and the chlorine reform as a colorless halide when the glass is no longer exposed to light, making it suitable for indoor-outdoor dark glasses. From this development came the breakthrough in glass ceramics using controlled nucleation and crystallization and the promise of improved strength and electrical insulation.

A photosensitive pigment suspension of silver chloride is an expensive way to darken window glass. As the intensity of the sunlight is increased gradually, so is the darkness of the windows due to the formation of particles of silver suspended in the glass. As the intensity of light decreases toward the end of the day, the transparency of the glass is restored by the reformation of invisible silver chloride. Future developments and expanded use depend on the cost and the demand for more comfortable living conditions.

8.6 Safety Glass

When broken, glass has a tendency to form long cracks and large fragments with razorlike edges, even when annealed (heat-treated to remove internal strains). Specially manufactured to avoid flying fragments, safety glass is made by introducing a wire or plastic composite or by tempering, thereby greatly reducing the size of the glass fragments.

8.6.1 Wired glass

Wired glass is a type of safety glass with a wire framework designed to reduce the danger of flying glass. A steel-wire mesh is embedded between two slabs of molten glass or rolled into a single strip of molten glass. Although this glass is no stronger than the same glass without a wire mesh, the wire not only retards the extension of cracks but holds the fragments together to keep them from flying into long, jagged slivers.

8.6.2 Tempered glass

Glass is tempered (toughened) by reheating it in its finished shape at 650°C (1202°F) until it becomes soft. By cooling rapidly both sides of the glass at the same time with jets of air or by immersion (quenching) in a bath of oil, both sides of the glass are given a permanent compressive stress without stressing the still fluid interior of the glass. The glass is rolled in and out of the furnace repeatedly until a uniform compressive stress is produced. The tempering process toughens the glass by three to five times—the compressive stress must be overcome by a strong external stress before brittle fractures can occur. In practice, the requirement that the interior of the glass sheet remain unstressed limits the sheet thickness to about 3 millimeters ($\frac{1}{8}$ inch).

Although the air jets set up a large temperature gradient from the surface of the glass to the still hot interior, any stresses set up as the glass cools are fully relieved. It is not until the surface approaches room temperature that most of the tempering stress occurs. At this point the air jets are turned off and the surface of the glass warms from the still hot interior. Thus the "frozen" surface layers of the glass try to expand as they heat up and produce the compressive stress in the surface, thereby balancing tensile stresses in the interior of the glass.

The finishing of the edges, all holes, and cutting to size must be done prior to the tempering process because this glass shatters into very small fragments if the surface is broken. If a crack is started, it spreads rapidly over the entire surface of the glass because of multiple branching at the tips of each crack. This causes the entire sheet to disintegrate into small, rounded fragments, which lack the sharp edges that can produce dangerous cuts. Applications include skylights, windows, sliding doors, tub and shower enclosures, and glass partitions.

Tempered glass should be used wherever people are likely to come into bodily contact with flat glass. Flat sheet glass is frequently found in older homes and buildings, where it has become more brittle with age and presents a potential hazard. At every opportunity this aged glass should be replaced with tempered glass because of the seriousness of bodily injury if it is accidentally broken (Fig. 8.1).

Tempered glass is reported to shatter spontaneously, but this is rare. Extensive research has provided only speculation as to the cause. Some of the more plausible reasons are the presence of impurities such as nickel sulfide "stones," faulty tempering, faulty glazing installations, accumulation of scratches, and excessive solar radiation stress.

The safety features of tempered glass easily outweigh any problems. Since a mass of tempered glass chips falling from a height could be dangerous, precautions should be considered. For example, if used

Figure 8.1 Aged annealed glass may shatter into sharp daggers when bumped, causing severe injuries. The shower door should open outward and be fitted with tempered glass.

above ground level in double-glazed (insulating glass) windows, tempered glass might be limited to use as the inside panel. Or the windows could be recessed to provide a ledge to catch any falling glass. Or the ground directly below could be a landscaped area rather than a busy passageway.

In this age of irresponsible pranks and terrorism, tempered glass appears in many variations. For example, if a stone or a bullet hits the windshield of an automobile or locomotive, it is essential that the glass permit the driver a clear view of the region ahead. For this reason the exterior glass sheet of the laminate is usually given a moderate tempering compared to the interior sheet because large glass fragments provide better visibility than small fragments of fully tempered glass. If the driver or front-seat passenger is thrown into a windshield, the fully tempered glass on the interior causes the inside sheet to crumble into small, harmless pieces.

8.6.3 Chemically toughened glass

The replacement of lithium ions in the surface of the glass with more bulky sodium ions, or even larger potassium ions, results in crowding or surface compression. The process appears to be an ion-exchange stuffing as a smaller ion in the glass surface is replaced by a more bulky ion in the interstitial solid solution the glass structure represents. Chemical tempering by ion exchange increases the strength of the glass by a factor of 2 over that of either annealed glass or that of polycarbonate (CR 39).

It is the practice for prescription eyewear to be chemically strengthened by an ion-exchange process. The lithium ions in the surface of the glass are replaced (ion exchange) by potassium ions by an immersion in molten potassium nitrate for 16 hours of treatment at 400°C (752°F). Its impact strength, as tested by dropping a 2.5-centimeter (1-inch) stainless-steel ball from 127 centimeters (50 inches) above the surface, showed the chemically toughened glass to be about fivefold stronger than thermally tempered glass. Although there is no change in the prescription shape of the glass lenses, there is no lower limit to the thickness of the glass that can be strengthened—as compared to glass tempering.

Although chemical strengthening of glass was favorably received for use in architectural panels and windshields of cars, it has been dropped because of its cost and the good performance of laminated tempered glass. Some of the obvious applications where safety overrules cost are the windows for airplane cockpits, space vehicles, racing cars, museum showcases, architectural panels, and commercial cookware.

8.6.4 Laminated glass

Laminated glass is a type of safety glass. It consists of a thin sheet of plastic between two sheets of thin glass. The sheet of plastic needs to be a clear and tear-resistant film, such as polyvinyl butyral, which is then heat-sealed under pressure between the glass sheets to form a unit.

The safety feature of embedded plastic in laminated glass is to reduce the danger of flying glass. The plastic clings tightly to the glass fragments. Laminated glass is frequently used for glass partitions, doors, or wherever glass comes into bodily contact with people.

Any kind of glass can be laminated, except glass with a raised pattern on its surface.

8.6.5 Special-purpose laminates

Acoustical glass is laminated with two or more thicknesses of plastic interlayers. Because of a high absorption of sound in the audible range, these laminated units are used as office partitions to reduce the noise level. They are used widely to soundproof radio and television booths and studios.

Burglar-resistant windows consist of dual glass sheets with an interlaying plastic sheet two or three times the usual thickness of butyral plastic to provide additional strength. This strengthened form of laminated glass serves as a deterrent to smash-and-dash thieves.

Bullet-resistant glass is used for the special security requirements at bank tellers' windows and ticket booths. This product has four or more layers of glass interlayered with plastic. The total thickness of the glass laminate ranges from 2.5 to 7.62 centimeters (1 to 3 inches) or more, depending on the type and caliber of gun the unit is designed to resist.

8.7 Insulated Glass

The continued growth of air conditioning in homes, stores, and buildings, along with the need to conserve energy, steadily increases the demand for insulated glass walls and windows. An insulated-glass window assembly consists of two sheets of glass separated by an air space and sealed together into a unit. The air between the two is confined in place by a sealant. The purpose of this air space is to reduce the flow of heat energy entering or leaving the building through the glass. By insulating the inner sheet of glass, this glass is kept from being chilled below the dew point of the air inside the building, thus preventing it from fogging over.

Different sealant systems are sometimes used, such as polyisobutylene mastic or a hot-melt butyl formulation for the primary seal because of their low moisture permeability. A more common sealant is the preformed butyl rubber used as a glazing tape because of its low moisture permeability. This tape is inserted into window frames. When the glass is fitted into place against the tape, its bulk and weight form a seal against the resilient butyl tape.

The need for better sealants, such as butyl rubber, occurred when the windows fell out of a 60-story skyscraper. The double-glazed windows originally had a clear pane on the inside, a reflective pane on the outside, a layer of air in between, and a lead-solder sealed frame separating the panes. The bonds between the reflective coating and the solder and between the coating and the glass fused the assembly into a single unit. Cracks developed in the solder through vibrations and extreme temperature changes, which transmitted the strains to the outer panes, causing the glass to break and fall out of the frames. Eventually, all 10,000 windows had to be replaced.

A more effective dual-sealant system uses polyurethane polyisobutylene or silicone polyisobutylene (silicone-PIB) sealants. Because of the excellent physical properties of silicone and its continual improvement, the silicone polyisobutylene system is often used as a typical double-seal system. The silicone provides the external seal around the edges of the two glass sheets, enclosing a layer of air between them. This acts as a moisture barrier as well as an adhesive to hold the edges of the sheets of glass together. The internal seal is formed by a bead (strip) which seals off the silicone-glass interface and acts as a backup moisture sealant. Building codes usually specify the silicone polyisobutylene system for structural-sealant glazing.

Although insulation increases with an increase in the thickness of the air layer as long as the air is relatively stationary between the sheets, there is an optimum thickness of about 13 millimeters (0.5 inch). A larger air space would allow some heat to be carried between the sheets by convection currents in addition to the normal heat transmission. Some manufacturers replace the air between the glass sheets with argon or sulfur hexafluoride because of their better insulating properties.

Moisture has a larger specific heat capacity than most other gases and carries more heat. Since its presence would decrease the insulation of the air layer, it is important to keep the air dry. Incorporation of a desiccant (drying agent), such as molecular sieves or dehydrated silica, in the space bar that separates the glass sheets can keep the air layer free of any moisture that gets past the sealants after a period of several decades of time.

One of the problems of insulated glass systems is "fogging" (condensation) of the interior surfaces of the glass assembly by the more volatile ingredients, especially when cost reduction is achieved by the use of inferior sealants or faulty installations.

Care must also be taken to ensure that the glass is securely held in its frame. A strong wind can create a problem by forcing the outer glass sheet to bow inward so much that it touches the inner glass. This can be avoided by pressurizing the air between the two sheets so that the outer sheet bows slightly away from the inner sheet. Another solution is to install a spacer between the two sheets to keep them from being blown together.

Another problem, due to the thermal contraction of large sheets of glass in cold weather, causes them to shrink away from their frames. Glass windows in tall buildings have been known to mysteriously pop out of their frames—aided, no doubt, by the slightly pressurized air circulating inside the building. This has been corrected by redesigning the metal frames holding the glass in place.

The all-glass (glass-edge) insulating unit is a variation of insulated glass. This airtight insulating unit has two glass sheets fused together with glass spacers around the perimeter. Although this all-glass unit should withstand the environment indefinitely, it lacks the elastomeric flexibility of the sealant system. The stresses set up by thermal expansion and contraction, wind, and earth movement usually limit its size and application to small windows in homes and commercial buildings.

8.8 Solar-Control Glass

Solar radiation passes through glass easily. Coming through windows, it overheats the interiors of buildings during the hot season of the year, causing an energy load on an air-conditioning system. During its multiple reflections around the interior of the building, this radiant energy is converted into thermal (heat) energy. This process is the familiar greenhouse effect. There are several methods being used to prevent this heat buildup and glare.

8.8.1 Tinted glass

Tinting window glass is another way to insulate the interior of a building from solar radiation and glare (Fig. 8.2). Glass can be body-tinted or surface-tinted.

Body-tinted glass is made by adding a pigment to the ingredients before the glass is melted. Ferrous oxide (FeO) gives the glass a light-green tint, and the tiny iron oxide particles suspended in the glass matrix act as a sun screen. Other tints available include the familiar gray or bronze. Some of the solar energy absorbed by the glass is emitted as radiation back outside the building and the rest goes into the interior of the building, partially defeating its purpose. To avoid breakage, the thermal expansion of the heated glass must be taken into account when securing the glass into the frame.

Figure 8.2 Tinted windows protect the interior of a skyscraper from the environment. A disadvantage is the reflection of heat and glare to the neighbors.

Gray and bronze body-tinted glass 6 millimeters (¼ inch) thick lets through about half the incident solar energy, including the same fraction of the visible light. By contrast, the same thickness of body-tinted green glass lets through only about half the solar energy and about three-fourths of the visible light. There is little difference in the absorption of solar energy. The amount of visible light absorbed depends on the color.

Surface-tinted reflective glass is tinted on the exterior surface and avoids absorption of solar energy by reflecting back the unwanted energy. This glass is made by coating the exterior surface with a thin deposit of metal or metal oxide. The reflective coating reduces the ambient radiation of energy into and out of the building.

The color and durability depend on the coating. Cobalt oxide and chromium coverings are reflective, durable, and can be placed on the outside of the glass. Less durable metallic coatings such as copper, aluminum, and nickel are only practical if used to coat the inside of the glass. Problems arise when the coating reflects so much light that the glass becomes a mirror reflecting everything around it, including nearby buildings and the blinding sun. And all this reflected heat and glare are not appreciated by close neighbors.

8.8.2 Coated glass

Another way to keep solar radiation from passing through the glass into the building is to use glass that is coated on one side with a film of indium tin oxide (ITO). The indium tin oxide coating efficiently reflects infrared radiation, but has little effect on the transmission of visible light. The coatings are available as films deposited directly on the glass or as a plastic film which can be laminated to glass. The films have good adherence, mechanical properties, and environmental durability. Developed for deicing aircraft windows, such coated glass also finds applications in solar energy cells.

8.9 Structural Glazing

The concept of using adhesives in structural glazing originated when accelerated weatherometer tests of silicone sealants exposed to water and ultraviolet light predicted their long-term performance. These results encouraged some leading engineers and architects to look on silicone sealants as glazing adhesives that could tolerate full exposure to the environment. After using silicone adhesives successfully to cement glass store fronts and first-floor windows in place, its use was expanded to windows in high-rise buildings.

In using a silicone sealant as the adhesive, the glass sheet is held in place by the adhesive without using brackets, and the interior of the building is protected from the environment.

Silicone sealants have been specially developed to meet specifications. The window can be a single glass unit or an insulating glass and can be a tinted or coated glass. The stresses set up by the pulsating winds are relaxed by the elastomeric properties of the silicone sealant. Success as an adhesive sealant has led to their expanded use in bonding anodized aluminum spacers to the glass, as well as to coated glass and to painted metal surfaces.

This technology requires skillful installation of the well-engineered materials into carefully designed exterior walls of a high-rise building. The simplicity of the installation, good performance, and reduced cost favor its growing usage.

A number of factors affect the performance of the window unit. The sealant bond must be strong enough to resist steady or pulsating wind pressure, the thermal expansion and contraction of the glass, and the deadweight load of the glass. Although the wind pressure is the largest factor, each factor is important and must be considered. The architectural design requires a careful analysis of these disruptive forces on the sealant and on the four edges of the glass sheet, which then determines the selection of the sealant modulus, its area of contact with the building mullion, and the thickness of the sealant bead (strip).

First tried in the 1960s, adhesive sealant silicones have performed well. Tests on commercial buildings show that the product retains two-thirds of its strength after 20 years of exposure. Extended use depends on how much longer it can withstand the weather, especially 282-kilometer-per-hour (175-mile-per-hour) winds to be expected in some regions. There is an enormous challenge and risk involved due to the difficulty of predicting the useful life of a sealant. It is prudent that the integrity of the bond strength be monitored closely over the life of the building.

8.10 Glass Fibers

The original glass insulating fibers were used in the building industry for thermal and acoustical insulation. The main source, mineral wool, was made by blowing air or steam through heated slag. These fibers are sharp on the ends and can be painful to the touch.

Now many tons of glass fibers are produced in the United States each year: two-thirds of it as blown fibers primarily used for thermal insulation purposes and one-third as continuous fibers for polymer reinforcement and related usage.

Owens-Corning Fiberglass Company pioneered the production of a fine continuous-filament glass fiber. This was soon followed by fine glass fibers for use in acoustics, heating, and air filtration. Each of these products starts with a glassy melt in a tank furnace (cupola) or in an electric furnace. Molten glass is forced through multiple orifices to make many continuous fibers at the same time. Glass is extruded either directly from a furnace in the direct-melt process or from the preformed glass marble process in which it is remelted before extrusion.

Blown fibers are produced by drops or streams of molten glass striking a jet of hot air to produce a fine cottony mass of assorted lengths and diameters. By using a clay or an organic binder, the fibers can be compressed into a mat. The mat owes its thermal insulating and acoustical properties to the multitude of tiny air pockets entrained in the glass mat.

The chemical composition of the glass used to prepare fibers is selected on the basis of melting point, bubble release, temperature working range, and ease of fiber formation. The composition of a typical glass insulating fiber is 50 percent silicon dioxide, 25 percent calcium oxide, 14 percent magnesium oxide, 10 percent aluminum oxide, and 1 percent ferrous oxide. Note the absence of sodium or potassium oxides and boron oxide.

A letter scale is used to indicate the diameter of a fiber, starting with A for the smallest and increasing alphabetically to the letter U. The common A size has a diameter of between 1.5 and 2.5 micrometers (millionths of a meter) or 1.6 and 2.7 millionths of a yard.

Continuously drawn fibers range in composition from soft glass to specialty types of glass used for cement fiber reinforcement, electrical insulation, and chemical resistance. Alkali-resistant glass fibers used for reinforced concrete have the general formula of 60 percent silicon dioxide, 15 percent sodium oxide, 3 percent potassium oxide, 5 percent calcium oxide, 7 percent titanium dioxide, and 10 percent zirconium oxide.

Glass fibers resist rotting, blistering, wrinkles, and cracking. They are woven into fire-retardant textiles for use in clothing, draperies, carpets, and wall coverings. Future developments and expanded uses are expected in fiberglass mats for bituminous built-up roofing membranes.

When using glass fibers we avoid the health hazards that often occur with some organic fibers and asbestos.

8.11 Glass Ceramics

Renewed interest in glass research followed the discovery, by scientists at Corning Glass Works, of methods to prepare new types of glass composites to improve their mechanical properties. The formation of a microcrystalline phase, described by some as a separation phase in a homogeneous glass phase, led to the term "phase separation."

Phase separation, as commonly used in glass technology and especially glass ceramics, means that a solid phase has precipitated to intermingle with the remaining liquid phase. This process is basic to glass ceramics.

When combining glass and ceramics, some of the best properties of each are obtained. By using a nucleating agent, such as finely divided titanium dioxide, and by a controlled heat treatment, the result is a 90 percent microcrystalline glass with tiny ceramic crystals embedded in the glass matrix.

Glass ceramics are produced with low thermal expansion to resist thermal shock or with a high coefficient of expansion to match those of the metals to which they are to be attached. A small amount of this microcrystalline phase is invisible to the eye, but serves as a reinforcing filler to strengthen the glass structure. In larger amounts this microcrystalline phase gives an attractive milky appearance due to the multiple reflections of light from the tiny crystal surfaces.

An industrial process consists of melting the glass ingredients with a nucleating agent, such as titanium oxide, and cooling the molten glass mixture in the shape of the finished object or article. Then reheating and controlled cooling produce nucleation and the desired amount of microcrystallization for the glass ceramic. With the wide variety of types of glasses available and the range of controlled nucleation agents possible, their thermal expansion can be varied over a wide range.

Thermal and mechanical shock resistance are further improved by the use of aluminum lithium silicon oxide glass. Fortunately, the failure by deformation and creep that occurs in metal does not occur in glass ceramics. Even the tendency of ceramics to fail in tension is offset by the glass matrix.

One of the main differences between this recently developed material and the usual ceramic is its improved properties. Because they are less porous, glass ceramics is more impervious to stains and moisture than are ceramics. In addition, the glass-ceramic composite is more shock-resistant because the cracks that would normally start at a grain boundary or an imperfection in a ceramic surface are arrested by the microcrystalline network of the glass structure.

The unique features of glass ceramics account for their extensive uses, ranging from oven cookware to the nose cones of rockets (Pyroceram, courtesy of our space research).

8.12 Recycling

Glass of all kinds amounts to about 7 percent of the solid waste found in our landfills. Some glass is recycled by industry, by remodelers, and by consumers.

Glass processing plants customarily recycle their defective material by sending it back through the melting process. As an example, preconsumer waste from lightbulb manufacturing processes can be turned into floor tiles made of 60 percent recycled glass. These beautiful tiles are used instead of conventional vinyl or ceramic tile.

Depending on community awareness and the convenient location of collection centers, much of the postconsumer waste glass that normally finds its way into landfills (bottles, jars, old windows, old shower doors) can be sorted and recycled into useful materials. Neither the clarity nor the composition is important when postconsumer glass is used in the making of fiberglass for thermal insulation. Still, the cost-effectiveness of recycling depends on the amount of energy required to collect, transport, and process the materials to be recycled.

Terms Introduced

Amorphous solids	Laminated glass
Annealed	Lead glass
Borosilicate glass	Lites
Chemically toughened glass	Optical glass
Coated glass	Photosensitive glass
Crystallization	Safety glass
Float glass	Silica structures
Glass ceramics	Soda-lime glass
Glass fibers	Solar control glass
Glass-transition temperature	Tempered glass
Glazing	Viscosity

Questions and Problems

8.1. Why is glass called an undercooled liquid?
8.2. What is nucleation and how is it different from crystallization?
8.3. What is meant by the term "metastable," and how does it apply to glass?
8.4. How does glass differ from a ceramic in structure?

8.5. What are the advantages of glass ceramics over a glass and a ceramic?

8.6. How does insulating glass conserve our energy resources?

8.7. Contrast the different types of safety glass.

8.8. What makes chemically toughened glass stronger than tempered glass, yet fracture differently than tempered or annealed glass?

8.9. Describe two types of colored glass.

8.10. What are some of the problems in recycling glass?

Suggestions for Further Reading

Harold Rawson, *Glasses and Their Applications,* Institute of Metals, London, 1991.

Horst Scholze, *Glass: Nature, Structure, and Properties,* Springer-Verlag, New York, 1991.

D. R. Uhlmann and N. J. Kreidl, eds., *Glass: Science and Technology,* vol. 5, Academic Press, New York, 1980.

Chapter 9

Protective and Decorative Finishes

9.1 Introduction	276
9.2 Film Formers	276
9.2.1 Alkyd Resins	277
9.2.2 Vinyl Resins	278
9.2.3 Polyurethane Resins	278
9.2.4 Acrylic Resins	279
9.2.5 Epoxy Resins	280
9.2.6 Silicone Resins	281
9.3 Paint Solvents	281
9.4 Pigments	282
9.5 Additives	282
9.6 Surface Preparation	283
9.7 Organic Coating Systems	285
9.7.1 Selecting the Right Paint	286
9.7.2 Prime Coat	287
9.7.3 Water-Emulsified Paints	287
9.7.4 Typical Paint System	288
9.8 Physical Properties of Paints	288
9.8.1 Solubility of Organic Substances	289
9.8.2 Solubility Parameters of Liquids	289
9.8.3 Solubility Parameters of Organic Polymers	289
9.8.4 Polar and Nonpolar Substances	290
9.8.5 Viscosity of Polymer Solutions	290
9.8.6 Rheology, Thixotropy, and Dilatancy	291
9.8.7 Gloss	291
9.8.8 Transition Temperature	292
9.8.9 Adhesion	292
9.9 Paint Problems and Correction	293
9.10 Lead Toxicity and Its Abatement	295
9.11 Health Hazards	296
Terms Introduced	296
Questions and Problems	297
Suggestions for Further Reading	297

9.1 Introduction

Coatings, both organic and inorganic, traditionally include paints, varnishes, and lacquers. Architectural coatings refer to those used on building materials and structures. These coatings can be protective, decorative, or both. Paints are usually pigmented drying oils or dispersions of resins in a liquid which, when spread over a surface, form a hard and durable coating. Varnishes lack pigments. Lacquers have cellulosic derivatives or other dissolved hard resins as binders that form a continuous coating on evaporation of the solvent.

Paint has three essential components. The first component is the film former (binder), which is usually a modified drying oil or a natural or synthetic polymer emulsified in water. The second component is a pigment, such as titanium dioxide. The third component consists of additives, such as biocides to prevent mildew, corrosion inhibitors, and plasticizers.

Essential to the performance of paint is an adequate surface preparation of the substrate on which it is to be applied. Such preparation often takes as much effort as the actual painting itself. Carefully detailed is the importance of surface preparation and the procedures that provide the basis for relating the chemical and physical properties of paints to their performance. The principles of adhesion are considered in depth.

Varnishes are effectively nonpigmented drying oil or emulsion-type paints. Lacquers, which initially used surplus nitrocellulose (guncotton) left over from World War I, are seldom used in architectural or industrial coatings other than cabinetry, having been largely replaced by more specialized coatings.

The environmental concerns of lead poisoning and solvent release to the atmosphere, along with the potential health hazards they represent, are carefully considered. Material safety data sheets (MSDSs) are discussed.

The chapter concludes with a discussion of the nature, cause, and cure of common paint problems that plague professional and amateur painters alike.

9.2 Film Formers

Organic film-forming resins are made in a wide variety of molecular structures, often named after the monomers used and the type of organic reaction used in their production. As an example of a film former, the polyesters are named after the reaction of an alcohol (R_1—OH), where R_1 represents the hydroxyl-containing component, and R_2 stands for the portion of the molecule containing the —COOH group of an organic acid (R_2—COOH) to form an ester with the molecular structure R_1CH2—O—CO—R_2. To illustrate, an alkyd is a special case of an ester, usually involving polymerization with a drying oil modifier.

9.2.1 Alkyd resins

Alkyd is a combination of the terms alcohol and acid. One early type of alkyd is made from a trifunctional alcohol such as glycerin and phthalic acid. It is known as Glyptal. This thermoset (cross-linked) polymer is formed by baking the coating in place. A second type is a linear (thermoplastic) resin formed by the reaction of a dihydroxyl alcohol, such as ethylene glycol, with phthalic acid anhydride. This alkyd resin is solvent-soluble and can be made into solvent-based paint. Fatty acids made from seeds and animal fats can be used in place of phthalic acid to form inexpensive resins. The third type of alkyd results from using a drying oil, such as linoleic or linolenic acid. It dries by reacting at its unsaturated carbon bonds with oxygen from the air. Alkyd-type coatings require the use of metallic dryer combinations which are highly complex and very stereospecific. These specialty alkyd resins are used for their good gloss, low cost, and durability.

Developed in the 1930s, modified alkyd resins retain about 20 percent of the architectural coating market in spite of the various synthetic coatings now available. Formulation of the alkyds has become a science as well as an art. As mentioned, alkyds are oil-modified or fatty-acid-modified resins that are made from polyalcohols and polybasic (polycarboxylic) acids. The oil can be saturated or unsaturated. Short-oil alkyds contain up to 40 percent oil. Medium oil contains from 41 to 60 percent oil, whereas long-oil alkyds contain 61 to 70 percent drying oil. Oil refers to the naturally occurring or modified polyhydroxyl esters of fatty acids, such as linoleic, linolenic, and palmitic acids.

Short-oil alkyds, when combined with urea resins, make industrial paints for metal and synthetic wood. Medium-oil resins are used for refinishing cars and trucks. Long-oil resins, such as those modified with soybean or linseed oil, are used to paint homes and commercial buildings and for corrosion-protection coatings.

Certain compatible combinations of alkyds with other resins are made by mixing the alkyd with a variety of other polymerized resins such as chlorinated rubber, poly(vinyl chloride), amino, phenolic, styrene, epoxy, silicones, and acrylics. For example, chlorinated rubber or poly(vinyl chloride) is used for corrosion protection. Short alkyds improve gloss, flexibility, and adhesion when added to cellulose nitrate. Heat resistance, gloss, and durability are improved by mixing alkyds with silicones. Improved adhesion to chalky surfaces by latex paint results from the addition of long-oil alkyds. The addition of acrylic resins or polystyrene to alkyds improves their hardness and shortens the drying time of alkyds.

Copolymerization occurs by the reaction of their alkyd hydroxyl or carboxyl groups or the double bonds of their fatty acids. For example, monomers of isocyanate or epoxy react with the hydroxyl alkyd groups. In another type of reaction, the alkyd carboxyl groups react with ethylene diamine and linoleic acid products to form thixotropic resins.

The wide variety of alkyds is further illustrated by the reaction of other dibasic acids such as succinic acid with trihydroxy alcohols such as glycerin. Thus, by increasing the number of reactive alcohol (hydroxy) groups from two to three, a cross-linked thermosetting resin is formed. Pentaerythritol, a four-hydroxy glycol, is used widely. Various fatty acids are obtained from natural oils and can be used to form numerous groups of alkyds. The list of useful oil-modified alkyds is practically unlimited.

9.2.2 Vinyl resins

Vinyl resins made by addition polymerization of vinyl monomers are easily applied in solvent or emulsion coatings. Poly(vinyl acetate) is one of the most popular vinyl polymers used as a paint film former. Vinyl emulsions are widely used in the construction industry. They are tasteless, odorless, color-stable, and flexible. In solution form those requirements are needed for their extensive use on food and drink containers, paper, and foil used for food products, as well as for coating wire. They include poly(vinyl acetate) and poly(vinyl chloride) made into lacquers or emulsions. Poly(vinyl acetate) is brittle and needs a plasticizer or is used as a copolymer.

9.2.3 Polyurethane resins

Polyurethane (isocyanate) resins are used to make elastomers, flexible fibers, flexible or rigid foams, as well as industrial coatings. The resin provides the paint with good abrasion resistance, adhesion, gloss, hardness, and chemical resistance. Some urethane alkyds and sunlight-resistant reactive urethanes are becoming more popular in architectural finishes.

These resins are formulated in one- and two-package paints. A common one-package alkyd-modified polyurethane paint is made from isocyanate reacted with a polyhydric alcohol, a drying oil, or a fatty acid. These alkyd-modified resins "dry" or form cross-linked polymeric coatings by reacting with oxygen in the air. The resulting resins have ester linkages of the alkyd type together with some urethane linkages (nitrogen-linked carbon chains). If drying oils have been used, the double bonds react with oxygen to further cross-link the polymer chains.

The two-package system is used to obtain faster drying, chemical resistance, flexibility, or thick coatings during application. One type of the two-package system consists of a liquid low polymeric form of a polyurethane paint containing unreacted isocyanate groups. The second part consists of an amine catalyst such as dimethyl ethanolamine. When mixed, the coating can be applied by dipping, spraying, or brushing. But like other two-package paints, there is a short pot (shelf) life before gelation. Popular two-package systems today consist of an isocyanate component which is mixed with a hydroxyl-terminated component prior to application or even during application. A longer pot life at room temperature is obtained with a phenolic-modified (phenolic-blocked) polyurethane coating, but it must be baked to unblock the polymerization reaction.

9.2.4 Acrylic resins

Acrylic resins are probably the most widely used resins in paints. These are thermoplastic polymers or copolymers of acrylic acid, methacrylic acid, their methyl alcohol esters, or acrylonitrile. Many monomers are polymerized readily by the use of ultraviolet light, heat, or a catalyst such as benzoyl peroxide. These common acrylic resins are hard and brittle and require the use of a plasticizer.

A large variety of modified acrylic resins is made solvent free (usually solid), as solutions in organic solvents, as water dispersions, or as water emulsions. Their properties range from the hard solids to fibrous or elastomeric structures. Their different physical and chemical properties are obtained by attaching a functional group such as the hydroxyl, the amine, or a carboxyl to the basic acrylic monomer.

There are two types of water-based polymers. One can be suspended as a colloidal dispersion in water; the other can be emulsified in water. By neutralizing the functional group that is attached to acrylic monomers after polymerization, salts, which ionize in water, are formed. These charged polymer ions repel one another and, thus, stabilize the dispersed polymer in water. By contrast, an emulsifying agent is required to make the acrylic water emulsion of high polymers. Note that only small-molecular-weight molecules or ions can be said to be dissolved to form true solutions; therefore, polymer molecules suspended in water are said to be dispersed.

Other variations of acrylic resins are possible by forming acrylic copolymers between acrylic and methacrylic monomers, or by making a copolymer between an acrylic monomer and styrene, acrylonitrile, or vinyltoluene. These copolymers have varied properties and unique applications.

For example, the copolymerization of monomers of acrylic acid and methacrylic acid promotes adhesion to metals; methyl methacrylates introduce hardness, weatherability, light stability, and gloss retention into the acrylic polymers; styrene increases resistance to detergent cleaners and salt spray, but reduces resistance to yellowing and gloss retention; and alkyl monomers add flexibility and hydrophobic properties.

9.2.5 Epoxy resins

Epoxy resins have several advantages over most other resins. For example, free hydroxyl in the polymer along the chains provides outstanding adhesion to metal and polar surfaces. The phenyl groups separate the chains enough to avoid crystallinity and brittleness, yet provide toughness. The linkages between carbon-chain segments give the resin good temperature stability. However, the phenol groups do absorb ultraviolet light, and the resulting deterioration limits their outdoor use to prime coatings for automobiles and prime and finishing coats for tanks, bridges, and other applications where chemical resistance and corrosion protection are more important than appearance.

Epoxy resins are usually made from epichlorohydrin and a dihydroxy alcohol or bisphenol A. The characteristic epoxy group has an oxygen atom joined to two adjacent carbon atoms to form a triangular-shaped ring. Addition polymerization frees one of the carbons for chain formation, while the other carbon forms a hydroxide group from the oxygen of the epoxide. Development of cycloaliphatic epoxies in place of phenyl-group substitution provides ultraviolet resistance with increased weatherability and flexibility.

Because of the reactivity of the epoxy and hydroxy groups, epoxy-coating technology has resulted in their use in high-solid low-solvent epoxy coatings such as water-emulsion and powder coatings. Specialty coating formulations made from copolymers, such as epoxy-polyester copolymers, provide extensive variations of physical properties in this versatile epoxy family of resins.

Common curing agents are amines like ethylene diamine, anhydrides, and other catalysts. Although the amine is volatile and causes serious skin irritation, this problem has been overcome by premixing part of the epoxy component with the amine component to form an adduct. The adduct reaction system forms a modified two-package system with a shelf life of 1 year or more at room temperature.

9.2.6 Silicone resins

Silicones are versatile synthetic polymers and are made with a wide range of physical forms and properties. They are characterized by good thermal and chemical stability, water repellence, weatherability, ultraviolet stability, but high cost. The basic silicone structure is a chain of alternating silicon and oxygen atoms with various organic groups attached to the silicon atoms.

Silicone resins used in paints are made from mixtures of methyl, phenyl, and methylphenyl dichlorosilanes or trichlorosilanes by a hydrolysis and condensation process. For example, methyl trichlorosilane in xylene is hydrolyzed by mixing it with water. The excess water containing the hydrochloric acid reaction product is separated from the xylene layer. The low-molecular-weight siloxane dissolved in the xylene is neutralized by washing with a water solution of sodium bicarbonate. Zinc octanoate catalyzes the condensation of any remaining reactant to form a high-polymer silicone resin.

Depending on the silane mixture used and the reaction conditions, the resulting silicone can vary from a soft polymer to a hard, flaky polymer. A large amount of methyl dichlorosiloxane in the reaction mixture produces a hard, brittle resin with poor low-temperature flexibility. By contrast, a high percentage of phenyl dichlorosilane in the reaction mixture makes a softer silicone resin with good high-temperature stability and a paint with better shelf life. Silicone-based paints find their applications in coatings for air-conditioning units, boiler stacks, and jet engines.

Silicone resins are blended with organic resins to reduce the cost of silicone resins and to overcome the relative softness of the silicones. When silicones are blended with organic resins such as alkyds, phenolic, acrylic, and epoxy, retention of many of the good properties of each is the result. These resins have a number of industrial and architectural applications.

9.3 Paint Solvents

Paint solvents can be classified as good (active) solvents, which dissolve (disperse) the polymeric film former by separating and, thus, solvating the polymer chains, or as poor solvents (diluents), which reduce the viscosity of paints by dilution. Aliphatic hydrocarbon solvents reduce the cost of a paint by replacing some of the more expensive active cosolvents. Since a diluent is a nonsolvent, it must not be used to such an extent that the polymer precipitates during application or drying of the paint. If evaporation of the more volatile active solvent leaves the diluent in too high a proportion during the drying of the paint, a milky or satin finish will result due to precipitation of the polymer. To ensure that a paint has a glossy finish, enough of the active solvent must remain to form a continuous film as the solvent evaporates. The solubility parameter and degree of hydrogen bonding of the solvent must match that of the polymer within permissible limits.

9.4 Pigments

Pigments include inorganic pigments, organic dyes, and extenders. A pigment blend is an insoluble solid which shields the substrate from ultraviolet light as well as providing a pleasing color and appearance. Pigment concentration is usually between 10 to 30 percent in gloss and up to 75 percent in mat finishes. Chalking occurs when the vehicle deteriorates because of ultraviolet light and leaves the pigment poorly coated by the paint vehicle (binder dispersion). Formation of loose surface material (chalking) that wipes off helps paint to shed dirt and discolorations and makes repainting easier. White pigments, such as zinc oxide and, especially, the rutile form of titanium dioxide, have good wetting properties and protection from ultraviolet light. The rutile form of titanium dioxide has the greatest hiding power (light-scattering power) of any white pigment because of its high refractive index. The usual destructive (photocatalytic) effect of ultraviolet light on resins is increased by ultraviolet backscattering. This is reduced by coating the rutile crystals with another metallic pigment, such as silica or zirconium.

9.5 Additives

Paint additives are materials used in small amounts (0.1 to 1.0 percent) to improve the properties of a coating. Additives often affect physical properties other than the ones for which they are selected. In addition, one additive may synergistically enhance the effect of another additive to produce a greater modification in a physical property than either one, alone. Some of the effects may be undesirable so that it becomes a tradeoff of benefits when selecting additives for a particular coating. Although much has been learned about how additives work, the optimum formulation is still an art as well as a science, especially since each application is slightly different.

Antiskinning agents inhibit the oxidation of the liquid surface of a paint or varnish by the oxygen in the surrounding air. An antioxidant evaporates from the surface once the wet paint film is formed, allowing it to skin over and dry. A common antiskinning agent used in alkyds is methyl ethyl ketoximine. Butyl aldoxine is an effective antioxidant and antiskinning agent in oleoresinous paints.

Antisettling agents retard the settling or agglomeration of pigments in the paint by forming a gel structure to confine them. Methyl cellulose (thickening agent) and bentonite-type clays (thixotropic thickening agents) are used in water-emulsion paints.

Leveling agents help paints to flow and avoid leaving brush marks. Often wetting agents are used to improve the leveling action.

Antifoaming agents are used in emulsion-type paints to prevent foaming and to obtain a uniform thickness of film.

Antifungal and antimildew agents reduce the surface discoloration of paints due to the action of fungi and mildew.

Antisagging agents are used to prevent a paint from running, sagging, or forming a curtain as it is being applied to a surface. Colloidal silica and silicates, such as bentonite clay, are antisagging agents which give the paint thixotropic properties (viscous solids that become fluids momentarily, when stirred). Castor oil or polyamides can be used to coat the suspended paint particles as they are formed into loose aggregates or clumps.

Coalescing agents help waterborne paint particles to flow, coalesce, and form a more continuous film as the water evaporates, especially at low temperatures. Alcohols, such as butyl Cellosolve and Carbitol, are used for this purpose.

Other additives are silicone surfactants used in trace amounts to coat dust and other particles that cause blemishes if not wet by the paint film when it covers the surface. Excess gloss can be prevented by adding to the paint a trace of silica or polyethylene wax.

Waterborne emulsion additives include paint-coalescing (active) solvents such as butyl acetate, biocide preservatives to reduce the threat of mold and bacterial degradation, thickeners such as hydroxy ethyl cellulose, surfactants such as lauryl sulfonate, and antifoaming agents.

9.6 Surface Preparation

One of the main reasons for performance failures, especially in adhesion, is inadequate surface preparation of the substrate to be coated. In general, the surface should be cleaned of all foreign material before painting. Additional treatment is required for adsorbent materials and metals. When repainting interior or exterior surfaces, the preparation can consume as much or more time than the final painting itself.

Adsorbent materials, such as composition board, should be waterproofed and sealed with a primer before painting with a finish coat. Deposits of dust, dirt, oil, grease, mildew, mold, efflorescence, and loose scale or rust must be removed from all surfaces to be painted. If the surface has not previously been painted, it should be in a cured stable condition and acclimated to the local humidity.

For example, some of the oil in green uncured lumber can bleed to the surface and cause discoloration or paint to peel. If green lumber must be used, a primer is needed to seal off the surface. Visual mildew or mold on previously painted surfaces is usually found in poorly ventilated areas with high humidity. To remove, scrub the surface with a 5 or 10 percent water solution of trisodium phosphate (TSP) containing a little added chlorine bleach. *Important:* Wear gloves, use a face mask, and work in a well-ventilated area. Rinse well and allow to dry before repainting.

Concrete surfaces are finished in such a variety of ways that freshly cured concrete can be difficult to paint. (Contracts for painting usually limit the guarantee period to 1 year.) New concrete should be aged for 1 year, or at least 6 months, before painting. In preparation for painting, the surface should be cautiously acid etched with dilute phosphoric acid and repeatedly rinsed until the acidity indicates a pH between a slightly acid 6.8 to a neutral 7.0, as shown by laboratory pH test paper. (Phosphoric acid is preferred to muriatic acid because the reaction product with lime is water insoluble. This reduces porosity.)

Wallboard (gypsum board), because of its porosity and tendency to soak up water and other fluids, can be plastered and then waterproofed and sealed with a primer that is formulated for this purpose. The surface preparation requires the usual cleaning and removal of loose sand and dirt.

Plastered cement exterior or interior surfaces of buildings should have all visible mildew removed and the area treated. Excessive chalk, dirt, or other foreign material should be removed by scraping or brushing. Defective caulking should be removed and replaced with a compatible sealant. Repair spall and concrete restoration with an epoxy mortar system. This should be followed with a prime coat.

Ferrous metals, when stored too long, can become rusty. Some structural steels, sheet metals, or building hardware are supplied with just enough protective coating to withstand average environmental exposure. Further protection with a primer or lacquer is usually needed for exposure to high humidity or salty environments.

Aluminum surfaces exposed to the weather need to be cleaned of any oil or grease and etched or roughened with a wire brush, an abrasive, or stainless-steel wool to provide surface irregularities for mechanical adhesion. Otherwise, the surface can be cleaned with an organic solvent and a commercial etching-type aluminum primer applied that contains zinc chromate for corrosion resistance, followed with a protective lacquer.

Galvanized surfaces, if unweathered, should be acid-etched with 5 percent acetic acid (household vinegar), thoroughly washed, and allowed to dry before painting.

Copper surfaces are usually allowed to corrode naturally to form the characteristic green copper carbonate finish since this coating is tightly adherent and protects the surface from further corrosion. However, if the original copper color is preferred, a similar treatment to that for aluminum can be used before applying a protective coating.

9.7 Organic Coating Systems

Paints are decorative as well as protective (functional) coatings. Architectural organic coatings include solvent-based paints and water-based paints (Table 9.1).

TABLE 9.1 Typical Architectural Paint Systems

Type	Binder	Thinner	Application
Stain/sealant	Acrylic	Water	Wood, hardboard, stucco, masonry
Primer/sealant	Vinyl-alkyd oil modified	Solvent	Plaster, wallboard, masonry
	100% acrylic	Water	Wood, hardboard, plaster, exterior/interior
	Vinyl-acrylic copolymer	Water	Drywall, plaster, exterior/interior
	Acrylic epoxy	Water	Concrete, stucco, plaster, masonry
Interior enamel	Acrylic latex	Water	Wood, drywall, plaster, masonry
	Acrylic latex	Water	Trim, windows, wood siding
Exterior enamel	Acrylic latex	Water	Wood trim, siding, metal
Exterior paint	Modified acrylic	Water	Stucco, concrete, masonry
Silicone/alkyd	Silicone/alkyd	Solvent	Durable exterior trim
Red oxide primer	Alkyd oil modified	Solvent	Ferrous metals; contains corrosion inhibitor

Methods of application of organic coatings include brushing, roller coating, dipping, and spraying. To avoid waste, electrostatic application techniques are used in some dipping and spraying processes. For example, the spray nozzle can have a positive charge and the object to be coated a negative static charge. Thus the liquid or powder being sprayed through the nozzle has a positive charge and is drawn to the surface to be coated by attraction between the opposite electrostatic charges.

Surface preparation is essential for good performance of organic coatings. With the possible exception of epoxy coatings, a prime coat is necessary to get the best results from the top coat. A prime coat applied to metal, wood, or cement is used to seal and protect the surface.

9.7.1 Selecting the right paint

In the selection of the best paint for a job it is important that the label on the paint can and the material safety data sheet (MSDS) be read carefully. Federal law requires that MSDSs be prepared by the paint manufacturer and be available for all company products for distribution to the purchaser on request. The MSDS, described in the first chapter, is a description of the safety precautions to be followed when using a product.

Modified alkyds have many excellent features, including flexibility, gloss, and good bonding with the substrate. In view of the emission standards of the EPA regarding volatile organic compounds, an advantage of alkyds is that, modified, they emit very low levels of volatile organic compounds as they cure. Also, the alkyd volatile organic compounds are not active smog formers, in contrast to the active smog-forming solvents required by the water-based emulsion paints for film formation.

The architectural paints sold to the public in retail outlets for home owners are mostly one-package water-emulsion (latex) acrylic paints. Their popularity is due to ease of application and cleanup. They have the additional advantage of avoiding the fire hazards of the previous organic solvent-based paints. (Emulsion cosolvents are more toxic in many cases.)

Organic solvents are used in paints containing a drying oil or in coatings such as shellac and lacquer. These organic solvents and diluents add to the cost, are a fire hazard, and in unventilated spaces pollute the air. Strict environmental controls on the emission of organic solvents in some parts of the country have caused these paints to be replaced by high-solid, low-solvent coating formulations of poor quality.

Solvent-type paints depend on the formation of a paint film by evaporation of the organic solvent into the air. Since volatile organic compounds are limited or banned by EPA regulations as air pollutants, organic solvent-type paints are largely replaced by water-based paints. These contain just enough organic solvent to produce a continuous paint film from the residue left by evaporation of the water.

9.7.2 Prime coat

For metals a corrosion-preventive prime coat bonds to the etched and roughened surface and provides a tightly adherent chemical layer for better bonding to the top coat. Resistance to environmental corrosion is also provided.

Wood is usually dry and porous, especially on the end grain. It can extract solvent from the finish coating, thus changing the formulation of the paint. Unprimed wood allows natural oils to come to the surface and solvents or water from the paint to raise the grain of the wood. The primer prevents this by sealing the surface.

Cement, when freshly poured, can benefit from a prime coat that retards the evaporation of water until it has cured. Silicate or organic sealants are needed when there is danger of moisture or salt migration through the cement wall or foundation. Migration of salts is not only unsightly, but indicates leaching of essential ingredients from the cement.

9.7.3 Water-emulsified paints

The two types of water-based paints are water-dispersed and water-emulsified, as discussed in Sec. 9.2.4. These were developed, in the case of emulsions, to make coatings with larger molecules than are possible with organic solvents. Replacing organic solvents with water reduces a potential fire hazard. In addition, the charge on the dispersed paint particles makes electrostatic dipping possible.

Water-based or latex paints contain a binder, pigment, active solvent, and a biocide prepared as a latex emulsion in water. The typical latex paint might contain 20 percent pigment, 40 percent binder, 2 percent coalescing (active) solvent, 25 percent water, and the rest additives. Thus water replaces the expensive solvents, makes the paint easier to apply, and reduces the air pollution caused by solvent evaporation. A biocide is needed in the water emulsion to protect the paint from decay. A fungicide is necessary to prevent mold from attacking the dried paint film. Mercury compounds (toxics) have been banned for this purpose by the EPA. Acrylics are usually resistant to biocidal action.

The suspended paint droplets are formulated so that a continuous polymer film results when the water has evaporated. The coalescence of the paint droplets remaining after the water has evaporated depends on the composition of the remaining organic solvents in the film. A problem arises if high relative humidity delays the evaporation of water, therefore allowing time for excessive evaporation of the organic solvents to occur before the droplets coalesce. This occurs as the highly concentrated polymer droplets coalesce to form a continuous polymer paint film. The amount of gloss in the paint depends on the amount of active solvent left in the droplets to form a smooth and continuous coating.

There is a critical humidity, air speed, and temperature limit which, if exceeded during the application of latex paint, is detrimental to its performance. This is especially important in hot and humid climates when dealing with porous or semiporous substrates.

9.7.4 Typical paint system

Table 9.2 gives the percentages of the ingredients in a typical acrylic latex paint that might be found on a label. Note that the labels list pigments (coloring agents) and vehicles. Vehicles include all the rest of the paint, such as solvent, resin, and additives such as wetting agents.

9.8 Physical Properties of Paints

Essential physical properties of paints are solubility in organic solvents and the ability to wet the surface, to be thin enough to spread over the surface and not run off, and to dry with the required amount of gloss. Each of these properties is important, but with the need of reducing the amount of organic solvent to be evaporated from paints

TABLE 9.2 Typical Acrylic Paint System

Ingredient	Weight, %	Purpose
Pigment		
Titanium dioxide	26	White pigment
Zinc oxide	5	White pigment and fungicide
Silicates/carbonates	3	Extender (reinforcing pigment)
Vehicle		
Soya alkyd resin	1	Gloss additive
Acrylic emulsion*	18	Resin binder
Water	47	Thinner

*50% acrylic resin and 50% water.

as they dry, their solvency is vital. This is even more important with latex paints, which contain just enough organic solvent to form a continous paint film as the water evaporates.

9.8.1 Solubility of organic substances

When two substances are soluble (miscible) in one another, the one in the majority is called the solvent and the other the solute. A prediction of the miscibility, or the extent one liquid dissolves in the other, can be made by comparing their solubility parameters (cohesive energy density).

9.8.2 Solubility parameters of liquids

The solubility parameter of a substance is defined as the square root of the ratio of the molar heat of vaporization to its molar volume. This parameter is a measure of the force of attraction between the molecules of a substance. Thus the better the match in solubility parameters, the more soluble the two substances will be in one another.

For example, the low end of the solubility parameter scale starts with aliphatic mineral spirits at 6.9. The solubility parameters for the aliphatic alcohols are in the range of 11.0 to 15.0 because of hydrogen bonding with the glycerine value at 16.5. The aliphatic esters range from 8.5 to 10.0, whereas the aldehydes and ketones have values of between 9.0 and 10.0. Nitrated organic solvents range between 10.0 and 13.0, whereas chlorinated solvents range between 9.0 and 10.0. The large hydrogen-oxygen force between water molecules gives water the highest solubility parameter of 21.4.

9.8.3 Solubility parameters of organic polymers

Since the film formers used in paint are usually nonvolatile solids, it is not easy to calculate their solubility parameters. The solubility parameters for resins are obtained by using the value of the solvent that best dissolves them, or by using as an approximation the value for their monomers. Thus the value of the parameter for a polymeric resin has a range of values depending on its preparation. For example, nitrocellulose has a range of values from 11.0 to 13.0.

For ethyl cellulose the solubility parameter ranges from 8 to 11, for poly(methyl methacrylate) (PMMA) the range is 9 to 13, for alkyds it is 7 to 13, and for epoxy resins 9 to 14. Two substances will be more soluble in one another the closer they match parameters. In practice, these values serve as guidelines only. The optimum value is usually the lower value increased by one-third of the range. The solubility parameter of the solvent should be maintained within a range of 1 unit below or 2 units above the optimum value throughout the evaporation process. The final selection of solvents must be tested by application of the paint with a spray gun or a paint brush.

9.8.4 Polar and nonpolar substances

A polar substance will be attracted to the oppositely charged terminal of an electrolysis cell. A nonpolar substance will not have an electric charge. A polar solvent has reactive groups, such as an amine or hydroxyl, and dissolves or disperses other polar substances. The more polar the substance, the higher is its solubility parameter value. Nonpolar solvents are required to dissolve nonpolar substances. Simply stated, like dissolves like. Ionic or functional groups that have strong attraction for one another make a molecule or radical polar. Saturated aliphatic hydrocarbons are at the nonpolar end of the list, whereas water, acids, and alcohols are at the polar end of the list.

9.8.5 Viscosity of polymer solutions

The viscosity of a polymer solution is indicated by how easy it is to stir. It is measured by the force required to rotate a paddle in a cup of the solution. Conversion to units of viscosity, called poise, can be calculated directly or obtained by matching it with a liquid of known viscosity. For comparison of viscosities, it is convenient to compare the rate of fall of a steel ball in standard tubes of the solution. For liquids of very low viscosity, the rate of rise of an air bubble introduced from a syringe can be used. In the first case a continuous film of paint is formed as the solvent escapes through the tangle of polymer chains into the air. In the emulsion type the paint droplets coalesce into a continuous film after evaporation of the water followed by further loss of solvent as the paint film hardens. In the powder-type application without solvent present the paint particles melt and coalesce to form a continuous paint film. This method is limited to objects that can withstand oven temperatures necessary to melt the powder so that it can flow together.

As expected, the viscosity of a solution increases with the amount of polymer present. For the same weight of polymer in a solvent, the viscosity increases as the molecular weight of the polymer increases and decreases as the solubility parameter match improves. Thus, a poor solvent with a poor match in parameters causes the polymer chains to cling together in balls with the excluded solvent forming a second phase.

9.8.6 Rheology, thixotropy, and dilatancy

Rheology is the study of the flow of liquids. Thixotropy is a property of liquids to become thinner or less viscous when stirred. This is due to the alignment of tiny crystals in the liquid, such as clay particles in water. The liquid recovers its firmness and the crystalline platelets resume their random orientation as soon as the stirring stops. This is a valuable property for either paints or adhesives when applied to vertical surfaces in order to avoid a sag or run.

Polymers, such as the polyamides, also provide a thixotropic property due to the weak interatomic forces, called hydrogen bonds, between chains. Hydrogen bonds are formed in the polymer chain between positively charged hydrogen atoms and negatively charged atoms of oxygen or nitrogen. Dilatancy is the reverse of thixotropy. If the viscosity of a liquid increases with the shear stress of stirring, it is called dilatancy. Breaking up of balls or chunks of polymer chains will provide more drag on one another with a resulting increase in viscosity. When the stirring stops, the chains regroup into balls and slide over one another with a lower viscosity. (This is a necessary property of the fluid used in the automatic transmissions of automobiles.)

Extenders are added to provide thixotropic properties that allow the paint to thin and flow when stirred or brushed over a surface. Then it thickens so that it will not run or drip after being applied. Barium and calcium carbonates and sulfates are used as well as diatomaceous earth (the siliceous skeletons of diatoms mined from deposits in the earth). Kaolin is typical of the silicate clays used as fillers, and talc can impart a soft, slippery property.

Thickeners added to reduce the run of paint can be divided into resinous thickeners for aliphatic solvent systems and thixotropic materials, such as bentonite, for use in more polar aromatic solvent paints.

9.8.7 Gloss

The gloss of a surface coating is due to the mirrorlike property of its smooth surface (Fig. 9.1). Any condition, or additive, that roughens the surface will produce a nonglossy finish. Additives, such as finely divided silica, will rise to the surface in clumps to produce roughness unless dispersed with a surfactant. A small amount of dispersed polyethylene wax will rise to the surface of a coating to produce a roughened surface. This breaks up the regular reflection of light from the surface giving it its gloss.

Figure 9.1 Lighted at night, glossy green paint makes the repainted MGM Grand Hotel, Las Vegas, Nev., shine like a neon sign. An epoxy polyamide primer, 4 mils thick, and an acrylic urethane system of two coats, each 3 mils thick, are expected to retain gloss for 20 years. (*Courtesy of MGM Grand Hotel and Theme Park and Dunn-Edwards Corporation.*)

9.8.8 Transition temperature

The temperature at which glass changes from a solid to a plastic consistency is called its transition temperature T_g. Amorphous materials such as glass and high polymers do not have a fixed melting point, like pure metallic elements, but they gradually soften over a range of temperatures above T_g. For high polymers such as the acrylics, T_g is related to the number of carbon atoms in the ester group. For example, methyl methacrylate with one carbon atom in the ester group has a higher T_g than butyl methacrylate with four carbon atoms in its ester group. Copolymers of these two monomers have an intermediate T_g, which depends on their ratio in the copolymer. T_g also increases with the degree of polymerization (number of monomer units in the polymer) and the extent of cross-linkage in the polymer.

9.8.9 Adhesion

The energy of the bond, or the force holding two dissimilar materials together, determines its adhesion. In order to measure adhesion, the test should measure the force or energy expended to separate one surface completely from the other. Many adhesion tests have been devised, all destructive, and with each there is a question about the location and completeness of separation. In an evaluation of such a test it is necessary to report the percentage of cohesive failure of the bond as well as the percentage of adhesive failure.

Cohesive failure is the term used when the failure occurs either within the adhesive or within the substrate of the surfaces being bonded. Adhesive failure occurs if it appears that the separation is at the interface between the adhesive and the substrate.

Because in practice an adhesion test results in a mixture of cohesive and adhesive failures, the author has suggested that the noncommittal term "hesion" is more appropriate. An alternative is to measure the percentage of the failure that is adhesive and the percentage that is cohesive.

In the case of paint adhesion, coatings need high adhesive strength. In addition it is desirable that the coatings have more cohesive strength than the substrate. However, the coating usually becomes more brittle as its cohesive strength is increased, so there is a limit to the improvement for a given coating.

9.9 Paint Problems and Correction

Some of the more common problems that occur in painting include the following:

1. *Blistering and peeling.* Both of these moisture-related problems are caused by moisture accumulating beneath the paint film. If the film is new and still elastic, the pressure of the confined moisture vapor can stretch the film into a blister. If the film is broken, water and, perhaps, a little oil will be squeezed out. If the film is old and brittle, it will crack and peel in order to release the moisture.

Peeling of paint on the ceilings and walls of the top floor of a building may be caused by a water leak in the roof or outside parapet. This requires an inspection of the roof and parapet walls and caulking or sealing of any leaks.

Peeling paint on window frames is probably due to ineffective caulking around the glass-frame structure. Replacement of old caulk with a fresh silicone sealant may be needed.

2. *Milky paint surface.* Blushing produces a white milky surface in the coating. It is caused by using low-boiling-point solvents that cool the coatings below the dew point (temperature when moisture condenses from the air) as the solvents evaporate. Moisture from the air is condensed on the film and precipitates the polymer, thus giving it a milky appearance.

3. *Popping.* Popping, the formation of pin holes in the surface of a coating as it dries, is usually due to an imbalance of solvents in relation to the ambient temperature so that a continuous film forms on the coating before the bulk of the lower-boiling-point solvent has

evaporated. This could also occur with well-formulated paint exposed to the sun in extremely hot weather. The premature film temporarily confines a vapor bubble that leaves a hole or blister in the final coating as the vapor escapes. A solution is to sand the surface lightly and recoat it with a well-balanced solvent paint.

4. *Adhesion failure.* Lack of adhesion, with peeling of the coating, is often due to improper surface preparation, as described earlier. Failure in adhesion can also result if painting is attempted over a hard, glossy incompatible coating. The new paint is incompatible either because it cannot wet the surface or because the solvent cannot penetrate the original glossy coating. The solution is to sand and generally roughen the surface to allow more penetration of the solvents into the coating and to provide surface crevices to give the coating mechanical footholds on which to cling; in other words, to provide more opportunity for mechanical adhesion by increasing the surface area.

5. *Water stains on ceilings.* Stains and flaking paint on the ceilings of top floors of buildings may be caused by rain leaking around the elevator shaft or the ventilating pipes sticking up into the roof. The solution is to locate the roof leaks and recaulk and seal all possible leaks.

6. *Rust stains on ceilings and walls.* Rust stains appearing on the ceilings and walls of exposed balconies are easily recognized by the light yellow to red discoloration due to the presence of iron salts. The cause is water carrying the carbon dioxide from the air (which deteriorates the concrete in its path) and oxygen (which corrodes the iron reinforcing bars along its path) as the moisture seeps through the concrete. Unless the concrete and plaster have fallen away, exposing a portion of the rebar, the corrosion may not be serious enough to require repairing the rebars. To correct the problem, locate the source of the leakage and make necessary repairs before sealing and repainting the ceiling.

In some cases the rebars might not have been embedded deeply enough in the concrete. This would permit moisture and air to reach the iron rebars from the immediate surface instead of coming in from an overhead leak. If rusting is in an advanced stage and the rebars are exposed, the correction process is more involved. In addition to cleaning away the crumbly concrete, all rust must be removed from the iron, and a protective corrosion-resistant coating should be applied. After the concrete and plaster have been restored, the surface should be sealed with a prime coat before repainting.

7. *Efflorescence.* Repeated formation of a thin, white powdery deposit on exterior surfaces of concrete or brick walls is due to water leaks that leach out the salts as the moisture migrates through the materials and finally escapes to the surface. As these salt solutions evaporate, whitish powdery deposits are left behind. Efflorescence can also occur on the interior plaster walls from external leaks. These persistent deposits are not only unsightly but can cause the plaster to deteriorate and crumble. The solution is to seek out and repair the external water leaks.

8. *Age and weathering.* On many older buildings numerous paint coatings have been applied on top of one another without proper surface preparation. Whether due to loss of adhesion over the years, or due to the shrinkage of the top coat as it dries, the paint pulls loose in patches from the substrate. The solution is to sandblast, waterblast, or scrape the coatings away to expose a firm substrate for repainting.

9. *Loose, flaky appearance.* The lack of surface integrity of a freshly applied coating can be due to extreme temperatures or humidity at the time of application. These extreme conditions can prevent the carefully planned and orderly release of solvents from the surface of the coating as it films over and starts to dry. The limiting weather conditions that are specified by the manufacturer of the paint must be followed to obtain the desired performance of the product.

9.10 Lead Toxicity and Its Abatement

Pliny the Elder, in the early Roman days, wrote about the toxicity of lead. Yet lead was thought to be so indispensable that its use continued. Lead compounds were first recognized as toxic by federal legislation in 1971 with Public Law 91-695, titled "The Lead Based Paint Poisoning Prevention Act." In 1973 the United States Congress limited the amount of lead in paint for residential uses to 0.5 percent by weight in the dried film. In 1978 the content of lead in paint was lowered to 0.06 percent. In 1994 lead-based paints were allowed to be used on bridges and nonresidential buildings.

As efforts were made to restrict the level of lead and eventually phase out all forms of lead in paint, it still lingers in paint in many capacities. Lead chromate (red lead) is used as a corrosion inhibitor on the steel structures of buildings and bridges. Basic lead carbonates, basic lead sulfate, and lead chromate are used as pigments. Lead resinates, lead linoleates, and lead naphthenate are used as dryers.

9.11 Health Hazards

Finding paints that contain low levels of hazardous substances can be a challenging experience, particularly for individuals with chemical sensitivities and anyone else working with paint on the job. Information about the chemical constituents of paints (pigments, brighteners, fillers, biocides which increase shelf life and inhibit microbial growth, drying agents, and solvents) should be found in the MSDS provided by manufacturers and available to consumers when purchasing paint (and other architectural products). OSHA lists hazardous materials. More information can be found in toxicology handbooks.

The seriousness of exposure to the many types of paint vapors depends not only on the concentration of the vapor in the air, but also on the length of exposure and the type of solvent used. Thus the threshold level value will depend on indoor ventilation. In general, the threshold level value is the concentration of the vapor which, if exceeded for a given period of time, is believed to be harmful to a person's health. These levels have been reduced over the years for most chemical vapors as experience showed that they were more toxic than expected and sometimes cumulative.

Many of the new chemical products introduced into commercial use are made from toxic materials. Some of these are unstable in certain environments and produce toxic vapors. It is expensive and time-consuming to test each new product fully, especially the long-term and synergistic effects with other chemicals. The research chemist learns by experience, "If you can smell it, you should take steps to avoid it."

Important: When working with paints and potentially toxic substances, always wear protective clothing and a face mask if spraying, and provide plenty of ventilation to the area.

Terms Introduced

Acrylic resins	Polymer glass transition
Alkyd resins	Polyurethane resins
Epoxy resins	Rheology
Gloss	Silicone resins
Latex emulsion	Solubility parameters
Organic coating systems	Thixotropy
Pigments	Vinyl resins
Prime coat	Water-emulsified paints
Solvents	

Questions and Problems

9.1. What are the components of a latex paint?

9.2. Compare the purposes of a prime coating for metals, wood, and cement.

9.3. Discuss the advantages and disadvantages of an epoxy paint.

9.4. Discuss the hazardous nature of the chemicals used in paints.

9.5. When considering paints to use, how can an individual find out what volatile organic compounds are contained in the paints?

Suggestions for Further Reading

Vincent M. Coluccio, ed., *Lead-Based Paint Hazards: Assessment and Management,* Van Nostrand Reinhold, New York, 1994.

Jan W. Gooch, *Lead-Based Paint Handbook,* Plenum, New York, 1993.

Sybil P. Parker, ed., *McGraw-Hill Dictionary of Scientific and Technical Terms,* 5th ed., McGraw-Hill, New York, 1994.

Hans K. Pulker, Sr., ed., *Wear and Corrosion Resistant Coatings by CVD and PVD,* John Wiley & Sons, New York, 1989.

R. C. Smith, *Materials of Construction,* McGraw-Hill, New York, 1979.

Dieter Stoye, ed., *Paints, Coatings and Solvents,* VCH Publishers, New York, 1993.

G. P. A. Turner, *Introduction to Paint Chemistry,* 2d ed., Chapman and Hall, London, 1980.

Chapter

10

Design, Interiors, Furnishings, and Equipment

10.1 Introduction	299
10.2 Computer-Aided Design	300
10.3 Indoor Air Quality	300
10.4 Lighting Design	302
10.5 Office Equipment and Home Appliances	303
10.5.1 Office Machines and Supplies	303
10.5.2 Refrigerators and Ozone	303
10.5.3 Saving Energy in the Home	303
10.5.4 Home Heating Appliances	304
10.5.5 Home Security	305
10.6 Upholstery and Drapery Fabrics	308
10.7 Wallcoverings	309
10.8 Furniture	310
10.9 Flooring	310
10.10 Carpets	312
10.11 Cleaning Agents	312
10.12 Outdoor Decks	313
10.13 What about Metrics?	313
Terms Introduced	313
Questions and Problems	314
Suggestions for Further Reading	315

10.1 Introduction

The technology is available for us to make choices that are earth-friendly, healthy, comfortable, and economical. The "greening" of products, called "eco-deco" by some environmentalists, is spreading to include home and office interiors, including furnishings, equipment, and decoration. Improving our environment is important for everyone, particularly for those people who are chemically sensitive. No one has to live with sick-building syndrome (SBS) or building-related illnesses (BRI).

Building codes in many localities are changing to reflect the technology that is now available. However, any building that is deemed "up to code" may not be good enough; any lower would be against the law. Our goal should be optimum safety, health, comfort, economy, and appearance.

The topics in this chapter include computer-aided design (CAD), warranties, suggestions for improving indoor air quality, lighting design, energy-efficient appliances, new sources of wood for furniture, flooring, wallcoverings, flammability of drapery, upholstery fabrics, and carpets, and cleaning agents.

By carefully checking product labels and the material safety data sheets (MSDSs) described in an earlier chapter, we can find cost-effective means of greatly improving the quality of our environment.

10.2 Computer-Aided Design

The manual tools (pencils, brushes, T squares) used by architects for generations to communicate ideas are common to design development. Now computer-aided design puts a new dimension onto the drawing board. Computer-aided design is fully integrated within the entire design process and allows architects, designers, and home owners (with the appropriate software) to instantly explore and develop ideas (to scale) in three dimensions (elevations, plans, sections). Colors, lights, and shadows can be added. Finish materials can be visualized. Costs are calculated instantly and construction documents issued. The design becomes a realistic look where nothing can be misinterpreted and anything can be easily changed. Anyone thinking of remodeling, renovating, or building from the ground up can obtain a "walk-through" of their plans by using computer-aided design.

10.3 Indoor Air Quality

People do not immediately die from air pollution, but they may suffer lingering effects: coughing, shortness of breath, eye, nose, and throat irritation, allergic reactions, rashes, fatigue, headaches, nausea, dizziness, asthma and bronchitis, permanent lung damage, and lung cancer. The control of outdoor pollution may be beyond our individual control; however, there is much we can do about minimizing indoor pollution.

During a 12-year series of studies by the EPA it was determined that the most dangerous exposure to air pollution is located indoors and is self-inflicted. Indoor air pollution is relatively independent of the surrounding outdoor pollution even if we are located in areas where the outdoor pollution is high. It is estimated that we spend about 90 percent of our time indoors, either in the home or at the workplace, where we may be exposed to high levels of hazardous pollutants.

A by-product of our unhealthy, built environment is the sick-building syndrome, a condition occurring in an estimated half of the office and government buildings, schools, and homes in the United States. The EPA estimates that 75 million office workers are at risk. Sick-building syndrome may be a combination of conditions contributing to poor indoor air quality: sealed buildings with a lack of incoming fresh air; development of molds, bacteria, and other toxins in air ducts; and a buildup of carbon dioxide, carbon monoxide, and volatile organic compounds released from building materials, finishes, furniture, adhesives, office equipment, and home appliances. A building is probably sick if a great many occupants report symptoms of sick-building syndrome, and those symptoms leave during weekends and vacations.

Building-related illness is a clinically defined sickness which can be traced to specific air-quality problems within a building: usually poor ventilation or inadequate servicing of heating and cooling systems. Building-related illness is sometimes a lasting and debilitating illness in which recovery does not occur during short absences from the offending building.

According to several studies, indoor air can be 100 times more polluted than outdoor air and costs approximately $60 billion a year in medical expenses and lost productivity.

The volatile organic compounds found in building products, such as carpeting, furniture, adhesives, and some paints, are formaldehyde and EPA-rated carcinogens such as benzene, toluene, and xylene. Exposure to these volatile organic compounds, outgassed over a period of many years, may cause short-term discomfort. Long-term exposure may include developmental defects, cancer, immune-system disorders, and other serious problems. Searching for and locating nontoxic or low toxic alternatives may be time-consuming, but safe (or safer) products are appearing on the market because of consumer demands and governmental regulations.

Often hotels, office buildings, and schools have improperly located fresh-air intake vents. It is important that the air intake vents be located away from loading docks and other exhaust systems. Vents (with filters) should be as high above street level as is practical (preferably on the rooftop) and away from air-vent exhausts and other nearby pollution sources. The American Society of Heating,

Refrigerating, and Air-Conditioning Engineers recommends a minimum of 0.56 cubic meter (20 cubic feet) per minute of fresh air per person. Many buildings provide half that much or less.

Reports indicate that the air quality in airliners may be hazardous to your health. Several studies show that economy class passengers get about 0.2 cubic meter (7 cubic feet) of fresh air per minute and first class may get 1.4 cubic meters (50 cubic feet) of fresh air per minute, whereas the pilots in the cockpit get 4 cubic meters (150 cubic feet) of fresh air per minute. Is it any wonder that you feel sick after returning home?

Scientists with the American Lung Association, the EPA, and other organizations remind us that the most important step toward eliminating indoor air pollutants is to ban the second-hand smoke of cigarettes, cigars, and pipes. These more than double a person's indoor exposure to hazardous particles and gases.

The toxic fumes of the solvent tetrachloroethylene, used in the dry-cleaning process, can be mostly eliminated by hanging clothing on a porch or near an open window with a fan blowing outward for at least 4 hours.

Humidity control, essential to maintaining a healthful indoor air quality, may prevent exposure to common allergens such as spores of mold and mildew, which increase during very humid weather. During extremely dry weather or in a heated environment during the winter, humidity should be increased by using humidifiers, pans of water on radiators, or other methods of adding moisture to the air. Otherwise dried and cracked mucous membranes become sources of infection.

10.4 Lighting Design

Personal comfort is increased, and significant cumulative electric energy savings are made when consideration is given to lighting design. First, natural daylight possibilities should be considered. Second, use white or light-colored furnishings and wall finishes to maximize the reflection of natural light. Third, use appropriate ambient lighting with three-way switches to adjust brightness, fluorescent lamps with state-of-the-art electronic ballasts, and high-efficiency triphosphor fluorescent lamps which produce natural colors. The triphosphor fluorescent bulbs are available everywhere. The initial cost is high, but they use 75 percent less energy, are more long-lived than the incandescent bulb, and produce a significant savings over time.

Two lighting controls which may add considerably to energy savings in offices and schools are occupancy sensors, which automatically turn off lights if no motion is detected after 6 minutes, and daylight dimming sensors, which adjust the ambient lighting according to the level of incoming daylight.

On the market is a window blind which has tiny perforations in the narrow slats. To control the light, this blind blocks out sunlight, bright lights, or glare while admitting soft daylight and some of the outside view.

10.5 Office Equipment and Home Appliances

Office equipment and home appliances are necessary to our lifestyle. However, they do present problems. Some consume enormous amounts of energy, some release chlorofluorocarbons affecting the ozone layer, there are electromagnetic fields, and chemicals escape into the air we breathe. Fortunately, there are immediate improvements we can make and other improvements as we recycle old equipment and appliances for new ones.

10.5.1 Office machines and supplies

Electromagnetic fields which emanate from all electric equipment, including copiers and computers, are suspected carcinogens. The local electric company can be called upon to check for electromagnetic fields and give advice on reducing or eliminating them. Inks, marking pens, fixatives, photographic chemicals, spray paints, cleaning agents, and adhesives emit volatile organic compounds and should be used in well-ventilated areas.

10.5.2 Refrigerators and ozone

Hydrochlorofluorocarbons (HCFC-141b) are a substitute for CFC-11, the blowing agent used to manufacture high-efficiency foam insulation in refrigeration appliances produced before 1994. By January 1, 1996, all refrigerators and freezers made in the United States must use non-CFC refrigerants. The reduction of CFC is expected to help restore our ozone layer and reduce the greenhouse effect.

10.5.3 Saving energy in the home

Much of the environmental pollution is caused by the extensive use of energy in the home. Fossil fuels used in electric generators are rapidly depleting our natural resources and their emissions pollute the air.

Nuclear reactors have become an international problem because of accidents and difficulty in disposing of spent atomic fuel.

Clothes washers are programmed to use less hot water. Most of the energy reduction is accomplished by using less hot water in the mix for a warm-wash setting and by substituting cold rinses for warm. To meet standards, wash options now are warm, cool, and cold.

Front-loading clothes washers require less water and less detergent. Clothes are lifted up and out of the water as the drum turns and receive a more gentle wash action and better cleaning.

The space-saving, labor-saving, front-loading combination washer-dryer (an all-in-one machine the size of one conventional dryer) is making a comeback in the United States after an absence of several years. Also, combination machines offer savings of raw materials in manufacturing and initial cost to consumers. European manufacturers offer several popular models; U.S. manufacturers have none as of this writing.

The Electric Power Research Institute and a consortium of electric utilities are testing microwave clothes dryers. This is not new technology; just a new application of existing technology. Anticipated advantages are 20 percent higher efficiency, shorter drying times, lower fabric temperatures (because microwaves heat only the water), and clothes do not shrink. Different settings can use only microwave power for delicate fabrics or combine microwave and the traditional heat method. Problems being worked out are potential hot spots with zippers and the wrinkle-removal feature for permanent press and synthetics, which requires more heat to relax the finish or fiber.

Considerable energy savings can be gained when gas or electric water heaters are set to the minimum temperature needed for dishwashers. If there is no dishwasher, the temperature should be set much lower to avoid an accidental scalding in the tub or shower.

A gas range should have an exhaust connection to the outdoors or a hood fan. Never use the open oven door for space heating; contaminants travel throughout the house. If a gas appliance is not in good working order and repair is too costly, consider a new model or an electric appliance instead.

10.5.4 Home heating appliances

If a kerosene heater must be used, burn low-sulfur fuel, fill the heater outside, and keep it clean and adjusted. It must be vented to the outside, and a source of fresh air should always be available indoors. Never use it overnight.

In a wood-burning stove use wood that has been split and dried for at least 6 months. Avoid smoldering fires, allow air to circulate in the firebox, provide fresh outside air to the indoors, and extinguish embers before leaving the home and at bedtime.

A natural gas furnace is the least polluting method of space heating when using fossil fuel. It does not contribute to sulfur dioxide emissions and reduces the contribution of nitrogen oxides to the air. Natural gas avoids most of the environmental problems associated with oil, coal, and nuclear energy.

Wind farms and solar voltaic cells may produce the energy sources of choice when they begin to compete economically with other methods of powering equipment.

10.5.5 Home security

Inexpensive small appliances to consider installing and using are smoke alarms, carbon monoxide detectors, and motion-detecting outdoor lights (Fig. 10.1). They provide safety and often reduce the cost of homeowner insurance. The smoke alarm will sound an alert about smoke and fire. The carbon monoxide detector will warn of invisible, odorless deadly fumes that can escape when combustion appliances incompletely burn fuel (kerosene, oil, gas, coal heater or furnace, gas range, gas water heater). The motion-detecting light is a convenience for the homeowner at night and a burglar deterrent. Dead-bolt locks installed on entrance doors may also reduce the cost of homeowner insurance. These suggestions are inexpensive and can be installed easily.

Safety and peace of mind can be found by using security bars attached to windows and doors (Fig. 10.2). They protect the house and occupants from intruders and can be worth the extra expense. With the installation of attractive and almost unnoticeable bars, windows can be left open (day or night), letting in cool breezes and providing security to the occupants.

Picture windows in homes, glass panes in doors, and show windows in stores can be protected inexpensively and without obstructing the view by placing a specially designed, transparent microthin film of plastic on the interior of the glass. This film creates a superstrong impenetrable barrier against crime, violent weather, and injury, and reduces the penetration of ultraviolet radiation, thereby increasing personal comfort and preventing the fading of fabrics. Designed to withstand an impact of up to 400 pounds, the film will prevent glass fragments from flying, as shown in Fig. 10.3.

Figure 10.1 Three home security measures. This vulnerable louvered outside door is protected from intruders by a motion-detecting light, dead-bolt locks, and security bars.

Figure 10.2 Sturdy decorative security bars over louvered-glass window prevent the entry of intruders and allow the window to be open for cooling breezes.

Figure 10.3 Microthin layer of film creates a superstrong impenetrable barrier against crime, violent weather, injury, and fading. (*Courtesy of 3M.*)

10.6 Upholstery and Drapery Fabrics

Hard surfaces reflect and intensify noise. The fabrics used in draperies, curtains, upholstery, and carpets absorb much of the sound reverberation caused by people operating office machines, home appliances, and other occurrences of everyday working and living.

When selecting window and upholstery fabrics for commercial buildings and residences, the most important consideration is the smoke-generation and the flame-spread index. There are no completely fireproof fabrics, except for glass fibers. Because the National Life Safety Code is not adopted in every locality, it may be necessary to contact the local fire marshal before using fabrics intended for public buildings and to submit fabric samples to a testing laboratory.

Some fabrics are rated as flame-resistant, meaning that they do not ignite readily and burn slowly if ignited. Natural fibers, such as wool and silk, and synthetics, such as nylon, olefins, and polyesters, are flame-resistant. In a lesser category are fire retardants, meaning that the combustible fabric is coated with a chemical to reduce or eliminate a tendency to burn. Most fabrics can be treated with fire-retardant finishes, which should last through 25 washings or 20 dry cleanings. Check the MSDS before using these treatments as the chemical sensitivity of the building's occupants should be considered.

Many fabrics have antifungal and permanent-press finishes which contain formaldehyde and other volatile organic compounds. Home draperies or curtains are easily made from washable dressmaking fabrics or sheets. Alternatives are shades or blinds, or the windows may be left bare if there is a nice view and enough privacy.

10.7 Wallcoverings

Wallpaper has been used for centuries. It was an inexpensive alternative to tapestries used to reduce drafts and the cold surfaces of stone walls in ancient buildings.

Natural wallcoverings may be made of grasscloth, linen or silk laminated to a paper backing, and cork. Lamination is costly. Felt or other heavy fabrics hung along walls may enhance the acoustical qualities of the walls. These coverings will withstand normal usage for many years.

Vinyl wallcoverings became available during the middle of the twentieth century. Each year vinyl is achieving increased usage in the interior-finish market as new and improved products are continually being developed. Their main features are beauty, durability, and low maintenance.

The vinyl wallcoverings consist of one or more layers of vinyl with a fabric, paper, or nonwoven backing. The backing adds strength and tear resistance to the finished product.

A vinyl compound generally consists of a poly(vinyl chloride) resin and a plasticizer, which is a combination of ingredients giving pliability and ease of processing to the vinyl resin. The chlorine in vinyl chloride makes it a fire retardant.

Plasticizers may give improved pliability at low temperatures, improved fire retardance, greater resistance to staining and abrasion, and better long-term aging characteristics.

Stabilizers are used as a processing aid to keep the product from turning yellow or brown under processing heats and light exposure.

Pigments must be heat-stable, lightfast, and nonreactive in the vinyl system. Quality pigments, especially reds and golds, are the most costly.

Generally, calcium carbonate is used as a filler to lower costs and to give certain characteristics to the finished product. Other additives may include flame retardants or fungicides to inhibit the growth of microorganisms.

Vinyl wallcoverings are produced by calendering, which consists of running the vinyl compound over a series of hot metal rollers to flatten the hot compound into a film or sheet. This film may be made and used separately, or it may be laminated under heat and pressure to a fabric backing. Sometimes the film is laminated to the fabric (using an adhesive) in a separate machine.

Special staph-resistant antimicrobial vinyl wallcoverings are made for hospitals, health facilities, and nursing homes. The antimicrobial additive, mixed with the vinyl coating when manufactured, is effective against the growth of mold, mildew, and bacterial organisms.

Most wallcoverings are made with vinyl cloth which emits chemicals along with the adhesive used to apply it. A few lines of all-paper or fabric wallpaper and some low-toxicity adhesives are on the market. Before a final decision on which wallcovering to use, it is best to ask the manufacturer or supplier for a copy of the MSDS to check the chemicals used against a list of agents with possible health risks. No product is produced which embodies all properties. Therefore, a balance of properties is the goal. Painting the walls, discussed in another chapter, is an option to consider.

10.8 Furniture

Some manufacturers are making an effort to find new sources of wood because ebony, rosewood, mahogany, lauan, koa, and teak are generally overexploited and scarce. Mature rubber trees, no longer productive, are used by some manufacturers. Also, North American cherry and mahogany from tropical rainforests are woods purchased through the Rainforest Alliance. This organization provides an opportunity to encourage sustainable forestry practices by encouraging the use of wood which is certified sustainable and grown by rainforest people.

Until plastics were developed, wood was the most common material used for furniture. Now there are many imitation wood-grain pieces of furniture on which a plastic veneer is placed over pressed wood. Avoid the volatile organic compounds of pressed-wood furniture whenever possible by checking the manufacturers' MSDSs before purchase.

Formaldehyde may be emitted from fiberboard shelving in closets and kitchen cabinets. Also, the plywood used may contain outgassing formaldehyde and adhesives.

10.9 Flooring

To avoid chemicals, the preferred flooring choices are ceramic tile, slate, cork tile, marble, or wood finished with low-toxicity sealants. Another option is natural linoleum made from linseed oil, cork powder, wood dust, and resins.

Studies indicate that hardwood oak and even soft pine will last a "lifetime" if they are properly cared for. A problem is that the materials protecting them, the sealers and waxes, are rated to last only 1 to 5 years. At that time traffic paths, scars, scratches, and stains can use some cleaning and repairing.

To clean deep stains, remove deep blemishes that do not respond to a light sanding by applying full-strength bleach and washing between repeated coats until the area becomes light gray. Wearing a dust mask, sand the entire floor, prime the dull spots, and then use steel wool to reduce any slightly raised grain.

Burn marks can be removed by scrubbing with a household cleanser or shaving off the burn with a sharp plane or chisel, being careful not to create a trench in the surface. Treat deeper burns as a deep-set stain, using bleach and stain primer. Open windows and use fans to improve ventilation.

Pet stains should be sanded and bleached. To remove the odor, try a proprietary mix sold in pet stores, or soak the stain with hydrogen peroxide and cover with a cloth soaked in ammonia until the stain fades or the solution dries. Ventilate the area. *Warning:* Never combine bleach and ammonia; they produce toxic fumes.

If you have open seams, test one or more wood fillers in an open crack or deep scar. After drying, apply a coat of stain to the filler and to some of the surrounding wood.

If you plan to sand the floors yourself, find a good library book to study or one at the hardware store where you can rent a sander. Make sure you have electric circuits of the right size at home to power a heavy-duty sander, which may draw 25 amperes and require a 30-ampere circuit. This is beyond the 15-ampere capacity of most household circuits.

After sanding and thoroughly vacuuming, apply a stain and wipe according to the manufacturer's directions. Begin with a minimum amount. It is easy to add more and let it soak in to give added color if desired.

To protect the finish, apply coats of wax, or at least two coats of polyurethane, which is available in a variety of sheens. Three coats are preferable, and the extra drying time (overnight) is worth the effort.

Important: When refinishing floors yourself, or when in the area while someone else is doing the work, be sure to wear a dust mask and provide plenty of ventilation to the area. After applying waxes or polyurethane finishes, keep the area open to ventilation and stay away until fumes have dissipated.

10.10 Carpets

Most carpeting comes with manufacturers' specifications, including flammability and smoke-density ratings by the American Society for Testing and Materials (ASTM). Consider these important ratings before making a purchase.

Wall-to-wall carpeting may cause health problems because of the chemicals used in the fibers, backing, pads, and adhesives. When buying new carpeting, look for the Carpet and Rug Institute indoor air quality label that indicates the product has passed tests for low emission levels of chemical pollutants. The U.S. Consumer Product Safety Commission receives many complaints of health problems after the installation of new carpeting in offices, schools, and homes because of chemicals outgassing from the carpet.

Other options to use include carpets with jute or recycled backing that is tacked down, not glued, and carpeting made from recycled plastic soda bottles.

While installation is going on, and for three days afterward, make sure there is plenty of ventilation. Then regularly use a vacuum cleaner.

When removing old and installing new carpeting, workers should wear a dust mask and ventilate the area for at least 3 days.

In the home, instead of vacuuming throw rugs, shake them outdoors. Use doormats at entrances to prevent tracking in dirt and contaminants from outdoor soil.

Dust mites, those microscopic creatures thriving in wall-to-wall carpeting, can cause allergic reactions. Since vacuuming does not eliminate them, nonslip and easily washed scatter rugs are recommended for sensitive people.

10.11 Cleaning Agents

Today the public is inundated by advertising claims and a confusing array of new "super-do" household cleaning compounds on the market. Directions should be followed exactly, but this is often difficult because of the small print, technical terms, and trade names used. Who really knows what the cleaning agents contain when the list of contents and the directions for use require a magnifying glass to read and a degree in chemistry to understand?

A wide selection of products is unneeded. Carefully determine a few basic cleaning agents for specific jobs. And, remember, never combine them.

Design, Interiors, Furnishings, and Equipment   313

10.12   Outdoor Decks

Most lumber for outdoor decks is treated with compounds containing arsenic to ward off mold and insects. (Refer to Chap. 6.) Finishes contain chemicals to prevent mold, mildew, and water penetration. These volatile organic compounds, unhealthy for individuals, may add to our greenhouse problems.

10.13   What about Metrics?

Because the Metrics Conversion Act of 1975 was voluntary, metrics never took hold. The United States, along with Liberia and Myanmar (formerly Burma), is one of only three countries in the world economy that is not on the metric system. Federal law requires that all plans and specifications for federally funded construction contracts be in metric units after October 1, 1996.

Since January 1, 1994, many technical reports have been in dual units, with metric as the primary units and the English equivalents in parentheses. On January 1, 1995, English units were dropped and all technical reports are required to use metric units.

The International System of Units is known as SI. Base units are the meter (m), kilogram (kg), second (s), ampere (A), kelvin (K), candela (cd), and mole (mol), and they are officially recommended for universal use.

Terms Introduced

American Society for Testing and Materials (ASTM)
Antimicrobial additive
Benzene
Building-related illness
Calendering
Carbon monoxide detector
Carcinogens
Carpet and Rug Institute
CFC-11
Chlorofluorocarbons
Combination washer-dryer
Computer-aided design (CAD)
Dust mites
Ebony
Electronic ballasts
Fiberboard shelving
Fire retardants
Flame-spread index
Formaldehyde
Fungicides
Grasscloth

Hydrochlorofluorocarbons (HCFC-141b)
Immune-system disorders
International System of Units (SI)
Koa
Lauan
Mahogany
Metrics Conversion Act of 1975
Microwave clothes dryer
Motion-detecting light
National Life Safety Code
Permanent-press finishes
Plastic veneer
Plasticizer
Polyurethane finishes
Rainforest Alliance
Recycled plastic soda bottles
Rosewood
Second-hand smoke
Security bars
Sick-building syndrome
Smoke alarms
Solar voltaic cells
Sulfur dioxide emissions
Teak
Tetrachloroethylene
Toluene
Triphosphor fluorescent bulbs
U.S. Consumer Product Safety Commission
Vinyl wallcoverings
Volatile organic compounds
Wallpaper
Wind farms
Xylene

Questions and Problems

10.1. List 10 methods of eliminating the sick-building syndrome or building-related illness.

10.2. What are the short-term and long-term disorders caused by volatile organic compounds?

10.3. Why are triphosphor fluorescent bulbs recommended?

10.4. What are the advantages of a combination washer-dryer (an all-in-one machine)?

10.5. Name two inexpensive, small alarms that should be in all homes or offices.

10.6. Define flame-resistant and fire-retardant. List fibers belonging in each category.

10.7. List the advantages of using computers in design.

Suggestions for Further Reading

Douglas R. Cochran and Kyle H. Hamada, "Design Tools: A Continuing Evolution," *Hawaii Architect,* Nov. 1993.

Matthew Gandy, *Recycling and Waste: An Exploration of Contemporary Environmental Policy,* Ashgate Publishing, Aldershot, England, 1993.

Herbert F. Lund, ed., *The McGraw-Hill Recycling Handbook,* McGraw-Hill, New York, 1993.

William F. Martin and Steven P. Levine, eds., *Protecting Personnel at Hazardous Waste Sites,* 2d ed., Butterworth-Heinemann, Boston, 1994.

National Audubon Society and Croxton Collaborative, Architects, *Audubon House: Building the Environmentally Responsible, Energy-Efficient Office,* John Wiley & Sons, New York, 1994.

Sybil P. Parker, ed., *McGraw-Hill Dictionary of Scientific and Technical Terms,* 5th ed., McGraw-Hill, New York, 1994.

William Rupp and Arnold Friedman, *Construction Materials for Interior Design: Principles of Structure and Properties of Materials,* Whitney Library of Design, New York, 1989.

Catherine Ann Solheim and Paulette Popovich Hill, "Home Economists as Environmentalists," *Journal of Home Economics,* vol. 86, no. 2, p. 20, Summer 1994.

Appendix A

Atomic Number, Mass, and Electron Configuration

Element	Symbol	Atomic number	Atomic mass*	1/K	2/L	3/M	4/N	5/O	6/P	7/Q
Hydrogen	H	1	1.008	1						
Helium	He	2	4.003	2						
Lithium	Li	3	6.941	2	1					
Beryllium	Be	4	9.012	2	2					
Boron	Bo	5	10.81	2	3					
Carbon	C	6	12.01	2	4					
Nitrogen	N	7	14.01	2	5					
Oxygen	O	8	16.00	2	6					
Fluorine	F	9	19.00	2	7					
Neon	Ne	10	20.18	2	8					
Sodium	Na	11	22.99	2	8	1				
Magnesium	Mg	12	24.31	2	8	2				
Aluminum	Al	13	26.98	2	8	3				
Silicon	Si	14	28.09	2	8	4				
Phosphorus	P	15	30.97	2	8	5				
Sulfur	S	16	32.06	2	8	6				
Chlorine	Cl	17	35.45	2	8	7				
Argon	A	18	39.95	2	8	8				
Potassium	K	19	39.10	2	8	8	1			
Calcium	Ca	20	40.08	2	8	8	2			
Scandium	Sc	21	44.96	2	8	9	2			
Titanium	Ti	22	47.90	2	8	10	2			
Vanadium	V	23	50.94	2	8	11	2			
Chromium	Cr	24	52.00	2	8	13	1			
Manganese	Mn	25	54.94	2	8	13	2			
Iron	Fe	26	55.85	2	8	14	2			
Cobalt	Co	27	58.93	2	8	15	2			
Nickel	Ni	28	58.70	2	8	16	2			
Copper	Cu	29	63.55	2	8	18	1			
Zinc	Zn	30	65.38	2	8	18	2			
Gallium	Ga	31	69.72	2	8	18	3			
Germanium	Ge	32	72.59	2	8	18	4			
Arsenic	As	33	74.92	2	8	18	5			
Selenium	Se	34	78.96	2	8	18	6			
Bromine	Br	35	79.90	2	8	18	7			
Krypton	Kr	36	83.80	2	8	18	8			
Rubidium	Rb	37	85.47	2	8	18	8	1		
Strontium	Sr	38	87.62	2	8	18	8	2		
Yttrium	Y	39	88.91	2	8	18	9	2		
Zirconium	Zr	40	91.22	2	8	18	10	2		

*Based on carbon-12.

Atomic Number, Mass, and Electron Configuration (*Continued*)

Element	Symbol	Atomic number	Atomic mass*	Electron shell						
				1/K	2/L	3/M	4/N	5/O	6/P	7/Q
Niobium	Nb	41	92.91	2	8	18	12	1		
Molybdenum	Mo	42	95.94	2	8	18	13	1		
Technetium	Tc	43	98.00	2	8	18	13	2		
Ruthenium	Ru	44	101.07	2	8	18	15	1		
Rhodium	Rh	45	102.91	2	8	18	16	1		
Palladium	Pd	46	106.40	2	8	18	18	0		
Silver	Ag	47	107.87	2	8	18	18	1		
Cadmium	Cd	48	112.40	2	8	18	18	2		
Indium	In	49	114.82	2	8	18	18	3		
Tin	Sn	50	188.69	2	8	18	18	4		
Antimony	Sb	51	121.75	2	8	18	18	5		
Tellurium	Te	52	127.60	2	8	18	18	6		
Iodine	I	53	126.90	2	8	18	18	7		
Xenon	Xe	54	131.30	2	8	18	18	8		
Cesium	Cs	55	132.91	2	8	18	18	8	1	
Barium	Ba	56	137.34	2	8	18	18	8	2	
Lanthanum	La	57	138.91	2	8	18	18	9	2	
Cerium	Ce	58	140.12	2	8	18	19	9	2	
Praseodymium	Pr	59	140.91	2	8	18	20	9	2	
Neodymium	Nd	60	144.24	2	8	18	22	8	2	
Promethium	Pm	61	(145)	2	8	18	23	8	2	
Samarium	Sm	62	150.40	2	8	18	24	8	2	
Europium	Eu	63	151.96	2	8	18	25	8	2	
Gadolinium	Gd	64	157.25	2	8	18	25	9	2	
Terbium	Tb	65	158.92	2	8	18	26	9	2	
Dysprosium	Dy	66	162.50	2	8	18	28	8	2	
Holmium	Ho	67	164.93	2	8	18	29	8	2	
Erbium	Er	68	167.26	2	8	18	30	8	2	
Thulium	Tm	69	168.93	2	8	18	31	8	2	
Ytterbium	Yb	70	173.04	2	8	18	32	8	2	
Lutetium	Lu	71	174.97	2	8	18	32	9	2	
Hafnium	Hf	72	178.49	2	8	18	32	10	2	
Tantalum	Ta	73	180.95	2	8	18	32	11	2	
Wolfram	W	74	183.85	2	8	18	32	12	2	
Rhenium	Re	75	186.20	2	8	18	32	13	2	
Osmium	Os	76	190.20	2	8	18	32	14	2	
Iridium	Ir	77	192.20	2	8	18	32	17	0	
Platinum	Pt	78	195.09	2	8	18	32	17	1	
Gold	Au	79	196.97	2	8	18	32	18	1	
Mercury	Hg	80	200.59	2	8	18	32	18	2	
Thallium	Tl	81	204.37	2	8	18	32	18	3	
Lead	Pb	82	207.19	2	8	18	32	18	4	
Bismuth	Bi	83	208.98	2	8	18	32	18	5	
Polonium	Po	84	(209)	2	8	18	32	18	6	
Astatine	At	85	(210)	2	8	18	32	18	7	
Radon	Rn	86	(222)	2	8	18	32	18	8	
Francium	Fr	87	(223)	2	8	18	32	18	8	1
Radium	Ra	88	226	2	8	18	32	18	8	2
Actinium	Ac	89	227	2	8	18	32	18	9	2

*Based on carbon-12.

Atomic Number, Mass, and Electron Configuration (*Continued*)

Element	Symbol	Atomic number	Atomic mass*	Electron shell						
				1/K	2/L	3/M	4/N	5/O	6/P	7/Q
Thorium	Th	90	232.03	2	8	18	32	18	10	2
Profactinium	Pa	91	231.03	2	8	18	32	20	9	2
Uranium	U	92	238.03	2	8	18	32	21	9	2
Neptunium	Np	93	237.05	2	8	18	32	22	9	2
Plutonium	Pu	94	(244)	2	8	18	32	24	8	2
Americium	Am	95	(243)	2	8	18	32	25	8	2
Curium	Cm	96	(247)	2	8	18	32	25	9	2
Berkelium	Bk	97	(247)	2	8	18	32	27	8	2
Californium	Cf	98	(251)	2	8	18	32	28	8	2
Einsteinium	Es	99	(252)	2	8	18	32	29	8	2
Fermium	Fm	100	(257)	2	8	18	32	30	8	2
Mendelevium	Md	101	(258)	2	8	18	32	31	8	2
Nobelium	No	102	(259)	2	8	18	32	32	8	2
Lawrencium	Lw	103	(262)	2	8	18	32	32	9	2
Rutherfordium	Rf	104	(261)	2	8	18	32	32	10	2
Hahnium	Ha	105	(262)	2	8	18	32	32	11	2
Seaborgium	Sg	106	(266)	2	8	18	32	32	12	2
Nielsborium	Ns	107	(262)	2	8	18	32	32	13	2
Hassium	Hs	108	(265)	2	8	18	32	32	14	2
Meitnerium	Mt	109	(266)	2	8	18	32	32	15	2

*Based on carbon-12.

Appendix B

Material Safety Data Sheets

Chemical manufacturers and distributors are required to provide material safety data sheets (MSDSs) to consumers and to put warning labels on their products. The MSDS on a product should provide information on the chemical and physical hazards of the material. However, many MSDSs are incomplete and lack accurate information. Trace amounts of chemicals are not required to be reported. The MSDS should be used as a guide only, and if more detailed information is needed, the manufacturer can be contacted directly. Compare several products before making a decision.

Sections of the MSDS and the information they should provide include:

1. Chemical identity and manufacture information
2. Hazardous ingredients and identity information
3. Physical and chemical characteristics
4. Fire and explosion hazard data
5. Reactivity data
6. Health hazard and medical treatment information
7. Precautions for safe handling and use
8. Control measures to avoid overexposure

Index

Acrylic adhesives, 207
Acrylic resins, 279
Addition polymers, 26
Adherence failure, 203
Adhesion:
  of paint, 292
  properties of, 203
Adhesive primers, 210
Adhesives:
  in structures, 206
  types of, 205
Aerosols, 222
Alkyd resins, 277
Allotropic forms of iron, 135
Alloys, 133
Alluvial soils, 32
Alpha iron, 136
Aluminum alloys, 144
American Society for Testing and Materials (ASTM), 312
Amorphous solids, 258
Annealing of metals, 142
Antimicrobial additives, 310
Architectural paint systems, 286
Asbestos fibers, 79
Asphalt, 221
  blown, 226
  cracked, 226
  cut-back, 226
  derivatives of, 225
  emulsions of, 226
Asphaltic bitumens, 225
Atomic number, mass, and electron configuration, 317
Atomic particles, 5
Atoms of elements, 5

Austenite, 131, 135, 137
Avogadro's number, 15

Bainite structure, 138
Batteries, 149
Beta iron, 131
Body-centered cubic crystals (bcc), 129
  (*See also* Metallic structures)
Building-related illness (BRI), 300

Carbon monoxide detectors, 305
Carcinogens, 301
Carpenter ants, 180, 181
Carpenter bees, 180, 181
Carpet and Rug Institute, 312
Cast iron, 136
Catalysts, 25
Cellulose, 175
Cement health hazards, 87
Cement symbols, 63
Cementite, 134, 137
Cementitious materials, 62
CERLA (Superfund), 46
Chemical bonding, types of:
  covalent bonds, 7
  ionic bonds, 7, 8
  metallic bonds, 123
Chemical equations, 16
Chemical equilibrium, 17
Chemical symbols, 19
Chlorofluorocarbons, 303
Clay masonry units, 104
Clay-mineral structure, 35
Cohesive failure, 203

323

324 Index

Colloidal-clay soils, 50
Colloidal systems, 222
Combination washer-dryer, 304
Computer-aided design (CAD), 300
Concrete, 67
   accelerating admixtures of, 72
   admixtures in, 70
   aggregates in, 69
   air-entrainment agents in, 73
   alkalinity of, 83
   durability of, 80
   efflorescence of, 86
   mineral admixtures in, 74
   porosity of, 69
   sulfating of, 82, 86
Concrete building blocks, 102
Concrete composites, 79
Copper alloys, 145
Coral building stones, 109
Corrosion, types of, 148
Corrosion protection, 85
Crystallization of glass, 256

Delta iron, 131
Dilatancy of mixtures, 204, 291
Displacement series of elements, 158
Double bonds, 21
Dust mites, 312

Ebony, 310
Efflorescence of cement and concrete, 86
Elastomeric materials, 194
Electron configuration of the elements, 317
Electronic ballasts for lamps, 302
Electrons, 5
Elements, 5
   table, 8
Emulsions of:
   oil in water, 221, 223
   water in oil, 221
Environmental hazards of masonry, 118
Environmental impact statement (EIS), 46
Environmental Protection Administration (EPA), 46
Epoxy resins, 27, 280
Eutectic formation, 132

Face-centered cubic crystals (fcc), 127
   (*See also* Metallic structures)
Ferrite iron, 135, 137
Ferrous metal structures, 131
Fiberboard shelving, 310
Fire:
   containment of, 247
   damage repair of, 248
   detection of, 247
   effect on concrete, 245
   effect on steel, 245
   effect on wood, 244
   effects at high temperatures, 243
   protection of materials from, 243
   protection of structural elements from, 248
Fire glass, 247
Fire retardants, 308
First law of thermodynamics, 3
Flame-resistant fabrics, 308
Flame-spread index, 308
Float glass process, 255
Fluorescent lamps, 302
Foamed plastic membranes, 227
Formaldehyde in fiberboard shelving, 310
Free radicals, 25
Fungi, damage in wood, 178

Galvanic corrosion of metals, 149
Gamma iron, 131, 136
Glass, chemical types of:
   borosilicate, 260
   lead, 260
   soda-lime, 259
Glass, film coatings on, 269
Glass, treatment of:
   annealed, 255
   chemically toughened, 264
   laminated, 264
   photosensitive, 261
   tempered, 261
   tinted, 267
Glass blocks, 111
Glass ceramics, 271
Glass fibers, 270
Glass lites (sheets), 265
Glass properties, 255
Glass recycling, 273
Glass structure, 257
Glass-transition temperature, 256

Granite, as a building material, 107
Grasscloth wallcoverings, 309
Grout masonry, 100
Gypsum plaster, 65

Half-cell concept, 150
Hesion failure, 203
Hexagonal close-packed crystals (hcp), 125
  (See also Metallic structures)
Humus, 49
Hydrochlorofluorocarbons (HCFC-141b), 303
Hydrogen ion concentration (pH), 17
Hydrolysis of salts, 18
Hydrophilic properties, 79
Hydrophilic surfaces, 204
Hydrophobic surfaces, 204

Immune-system disorders, 301
Indoor air pollution, 301
Insulated glass, 265
International System of Units (SI), 313
Interstitial solid solution, 130
Iron, allotropic forms of:
  alpha, 131
  beta, 131
  delta, 131
  gamma, 131
Iron alloys, 136
Iron carbide, 136
Iron-carbon phase diagram, 134
Iron ore smelting, 134

Koa wood, 310

Laminated glass, 264
Landscaping, contributions to environment, 57
Latex adhesives, 205
  (See also Adhesives)
Latex emulsion paints, 287
  (See also Paint)
Lauan wood, 310
Lava rock, 109
Lead alloys, 146
Lead toxicity, 295
Light, motion-detecting, 305

Lime, setting of, 66
Limestone, 107

Mahogany, 310
Marble, 109
Martensite formation, 138
Masonry, 98
  cement used in, 98
  hazardous practices, 116
  mortar used in, 95
  prevention of problems, 117
Masonry, treatment:
  cleaning of, 113
  restoration of, 112
Masonry units, 101
Material safety data sheet (MSDS), 45, 321
Metal cladding, 161
Metal composites, 146
Metallic bonding, 123
Metallic compounds, 133
Metallic corrosion, 147
Metallic structures, 124
  body-centered cubic (bcc), 129
  face-centered cubic (fcc), 127
  hexagonal close-packed (hcp), 125
Metrics Conversion Act of 1975, 313
Microwave clothes dryers, 304
Monomer unit, 24
Mortar, 95
  admixtures in, 99
  aggregates in, 98
  blended cement in, 97
  ingredients of, 97
  thickness of, 100
  water content in, 99
  workability of, 98

National Life Safety Code, 308
Net ionic equations, 16
Nitrogen fertilizers, 51
Nonferrous alloys, 143

Occupational Safety and Health Administration (OSHA), 45
Optical glass, 260
Organic materials, 19

Paint, 276
  acrylics, 288
  coating systems of, 285
  correction of problems, 293
  pigments in, 282
  prime coat of, 287
  solvents in, 281
  surface preparation, 283
  water emulsions of, 287
Paint additives, 282
Paint film formers, 276
Paint health hazards, 296
Paint properties, 288
  adhesion, 292
  gloss, 291
  hesion, 293
  rheology, 291
Paint selection, 286
Paper, 186
  manufacturing of, 186
  recycling of, 212
  self-destruction of, 187
Particleboard, 177
Passive steel, 83
Pearlite structure, 138
Periodic table, 12
Permanent-press finishes, 309
Pesticides, environmental hazards of, 55
Phase diagrams, 132
Phosphate fertilizers, 52
Photosensitive glass, 261
Piles and caissons, 41
Plant nutrients, 51
Plaster, 112
Plaster of paris, 65
Plastic building materials, 198
Plastic veneer, 310
Plasticizers, 309
Plastics, 198
  limitations of, 200
  recycling of, 212
  reinforcement of, 207
  residential uses of, 201
Plywood, 176
Polar and nonpolar surfaces, 290
Polybutadiene rubber, 202
Polychlorinated biphenyls (PCBs), 47
Polymer concrete, 78
Polymer fibers, 194
Polymer physical properties:
  flammability, 197

Polymer physical properties (*Cont.*):
  glass-transition temperature, 192
  plasticization, 193
  solution viscosity, 290
Polymer solvents, 193
  (*See also* Solubility parameters)
Polymer-transition temperature, 192
Polymeric units, 25
Polymers, 187
  natural, 188
  synthetic, 188
Polysteel forms, 246
Polyurethane finishes, 311
Polyurethane resins, 278
Portland cement, 63
  hydration, 65
Potassium fertilizers, 52
Pozzolanic materials, 97
Prestressed concrete, 78

Quenching of alloys, 142
Quicklime, 66

Rainforest Alliance, 310
Rebar corrosion, 83
Rebar passivity, 83
Rebars, 74
Recycling:
  of glass, 273
  of plastic soda bottles, 312
  of roofing materials, 234
  of used building materials, 44
Redox equations, 157
Reinforced concrete, 74
Residential roofs, 228
Rheology, 204, 291
  of soils, 34
Roof insulation, 230
Roof membrane materials, 226
Roof problems, 231
Roof warranties, 234
Roofing, 218
Rubber, types of, 202

Sacrificial anodes, 151
Safety glass, 261
  (*See also* Tempered glass)
Salt corrosion, 81

Sandstone, 109
Seaborg, Glenn T., 12
Sealant caulks, 238
Sealant primers, 237
Sealants, adhesive properties of:
 hardness, 237
 movement capability, 236
Sealants, types of:
 acrylics, 240
 butyl, 239
 polysulfide, 239
 silicone, 242
 urethane, 241
Second law of thermodynamics, 3
Secondhand smoke, 302
Security bars, 305
Shakes and shingles, 228
Shingles, types of:
 asphalt, 229
 mineral-fiber, 230
 wood, 230
Sick-building syndrome (SBS), 300
Silica structure, 257
Silicone resins, 281
Single-ply membranes, 226
Site investigation, 33
Site preparation, 37
Slaked (hydrated) lime, 66
Slate roofs, 109
Slump test, 73
Smoke alarms, 305
Soil compaction, 40
Soil mechanics, 34
Soil nutrients, major, 51
Soil profiles, 33
Soil texture, 32
Solar control glass, 267
Solar voltaic cells, 305
Solid solution formation, 133
Solubility parameters, 289
Spiders, 180
Stainless steels, 139
Standard electrode voltages, 153
Steel, heat treatment of, 141
Stone masonry, 92, 107
Structural glazing, 269
Substitutional solid solutions, 130
Surface tension, 204

Teak furniture, 310
Tempered glass, 261
Tempering of steel, 142
Termites, 180, 182
 control of, 184
Thermoplastic polymers, 25
Thermosetting polymers, 27
Thixotropic clay in soils, 35
Thixotropy of mixtures, 35, 204, 291
Tile masonry, 106
Tinted glass, 267
Trace soil nutrients, 53
Travertine, 107
Triphosphor fluorescent bulbs, 302

Universal Building Code (UBC), 2
Unsaturated hydrocarbons, 21
U.S. Consumer Product Safety Commission, 312

Vinyl monomers, 25
Vinyl resins, 278
Vinyl wallcoverings, 309
Viscosity of liquids, 255
Volatile organic compounds, 301

Wallpaper, 309
Water-cement ratio, 68, 99
Water-emulsified paint, 287
Water reducing agents, 70
Wattle and daub walls, 92
Weak acids, 18
Weathering of rocks and soils, 32
Whitewash, 65
Wind farms, 305
Wood:
 classification of, 175
 destructive organisms of, 178
 ingredients of, 175
 insect destruction of, 180
 recycling of, 212
 rot of, 179
Wood composites, 176
Work-hardening of metals, 141
Wrought iron, 136
Wythe masonry unit, 102

ABOUT THE AUTHORS

L. REED BRANTLEY is Professor Emeritus of Chemistry at the University of Hawaii at Manoa, and was formerly Professor and Chairman of the Department of Chemistry at Occidental College in Los Angeles. Dr. Brantley has conducted research on adhesion, water vapor barriers in buildings, and material coatings, and was instrumental in developing course materials on this subject for the University of Hawaii.

RUTH T. BRANTLEY was most recently an Associate Professor in the Department of Home Economics at the University of Hawaii College of Tropical Agriculture and Human Resources. She specializes in housing, home furnishings, interior design, and appliance technology.